Energy Management
and
Control Systems
Handbook

Second Edition

Energy Management and Control Systems Handbook

Second Edition

F. William Payne
and
John J. McGowan, CEM

Published by
THE FAIRMONT PRESS, INC.
700 Indian Trail
Lilburn, GA 30247

Library of Congress Cataloging-in-Publication Data

Payne, F. William, 1924-
  Energy management and control systems handbook.

Bibliography: p.
Includes index.
1. Buildings--Energy conservation--Handbooks, manuals, etc. 2. Heating--Control--Handbooks, manuals, etc. 3. Air conditioning--Control--Handbooks, manuals, etc.
  I. McGowan, John J., 1950-   II. Title.
  TJ163.5.B84P39   1988   658.2'6   86-46132
  ISBN 0-88173-028-9

Published by The Fairmont Press, Inc.
700 Indian Trail
Lilburn, GA 30247

Printed in the United States of America

10 9 8 7 6 5 4 3 2 1
ISBN 0-88173-028-9   FP
ISBN 0-13-277286-8   PH

While every effort is made to provide dependable information, the publisher, authors, and editors cannot be held responsible for any errors or omissions.

Distributed by Prentice Hall
A division of Simon & Schuster
Englewood Cliffs, NJ 07632

Prentice-Hall International (UK) Limited, London
Prentice-Hall of Australia Pty. Limited, Sydney
Prentice-Hall Canada Inc., Toronto
Prentice-Hall Hispanoamericana, S.A., Mexico
Prentice-Hall of India Private Limited, New Delhi
Prentice-Hall of Japan, Inc., Tokyo
Simon & Schuster Asia Pte. Ltd., Singapore
Editora Prentice-Hall do Brasil, Ltda., Rio de Janeiro

Preface to Second Edition

The original *Energy Management and Control Systems Handbook*, published in 1984, was a complete source of basic information on computer control systems. This field, however, is expanding, driving ahead, more rapidly than any other practical technology. To keep up with the changes, a comprehensive revision was absolutely necessary.

Further, the mood of the controls industry, and the marketplace itself, demands additional information on a whole different series of subjects. This Second Edition includes new data on Direct Digital Control, Automation System Program Management, Non-Energy-Related Benefits of Systems Integration, Information Management, Distributed Processing, and (perhaps the hottest topic of the '80's) Standardization and Remote Communication.

The editorial support of a co-author with hands-on experience as a Corporate Energy Manager, Energy Consultant, EMC Installer, and Controls System Specialist has been enlisted in order to bring the *Energy Management and Control Systems Handbook* completely up to date.

As a result, this Second Edition is a state-of-the-art guide to building automation and energy management systems that will enable consultants, engineers, and contractors to grasp the complete picture of the latest advances in this fast-moving technology.

F. William Payne
John J. McGowan, CEM

Contents

Contents

Contents

Contents

Contents

1

Automation Through Energy Management and Control Systems — Where We Stand Today

Bill Gerken,
Office of Building Energy
Research and Development,
U.S. Department of Energy

Automation through computerized control has become the standard in sophisticated building design. Widespread use of the technology resulted from the energy crises of the 1970's. Sustained use of these systems is due to the combined benefits of energy and cost savings, enhanced building control and information management. In this second edition of the *Energy Management and Control Systems Handbook,* emphasis will be placed on each of these areas of benefit. It is important to discuss each area in relation to the effect that technology advances have had, and to address some key lessons that the industry has learned through tracking system performance.

In the area of cost savings, research sponsored by the U.S. Department of Energy indicates that it is possible to save from 30% to 50% of the energy now being used in existing buildings, and to build new buildings that require from 50% to 75% less energy than those now standing. Both private and Federal studies have highlighted the opportunities for improving the efficient use of energy buildings through the development and introduction of advanced, energy conserving controls systems.

Though the emphasis of early system use was on sophisticated control and the energy savings to be achieved, a wealth of unexpected benefits have been realized through the presence of computerized control in buildings. It has been surprising that the value of enhanced control, information and communications can easily justify an automation system for a building owner, even without energy savings. With the remote communications feature, managers can ensure a management presence in facilities located far away.

There are a number of current and emerging issues which must be addressed if building energy controls are to fulfill their potential role in national energy conservation. This introductory chapter gives a better understanding of these issues.

EMCS INDUSTRY AND MARKET OVERVIEW

The Energy Management and Control System industry as a whole is both concentrated and competitive. Five firms garner over 80% of the industry's annual sales volume. Two companies dominate the market, although smaller firms produce the majority of electronic controls. The field appears relatively easy to enter, as suggested by the number of firms that do so each year. Balancing this influx is a substantial exodus, indicating that the industry may be an equally difficult one in which to remain. Among the reasons advanced for the latter condition are:

- many recent entries lack established nationwide distributor and/or service networks; and

- many new competitors seem to be product oriented, whereas a number of the successful, established firms are systems oriented, meaning that the latter are able to be more responsive to the needs of design engineers, contractors, and building owners/operators.

EMCS controls manufacturers individually range from conservative to progressive in their attitudes toward new technology. The conservative viewpoint has its historical basis in the need for reliability over a long service life with little maintenance and often untrained operating personnel, and in past low energy prices. More recently, an innovative thrust has evolved from the industry's traditional high level of competence, in combination with its ability to utilize, for building systems, technology originally developed for critical manufacturing process controls and aerospace applications.

The innovative thrust is not without its own challenges, however. Process controls are an order of magnitude higher than building controls in both equipment reliability and cost. To date, EMCS user satisfaction reports underscore the interrelationships between these two factors, by reflecting industry's only partial success in transferring reliability from one type of control to another while remaining cost competitive in the market.

It should be noted that controls technology is in a stage of rapid development and has not yet reached a plateau. Numerous generations of controls equipment have been introduced since 1970. Similarly, innovation in the semiconductor industry has probably not yet run its course. Further breakthroughs and subsequent cost decreases are still likely, and the controls industry will undoubtedly benefit from advances in these areas. Decreases in solid-state component costs affect only a minor portion of total controls system costs, however. Installation is a major cost element, and the increasing complexity of controls, as well as rising materials and labor costs, may prove to offset any further dramatic reductions in system costs.

Some people in the industry note that further acceleration of product introduction might prove counterproductive. This view is substantiated by the existing—and possibly increasing—knowledge gap between controls manufacturers and most of the remainder of the controls interest community. Unless this gap can be narrowed appreciably, the adverse effects—in terms of design consultant, equipment installer, and building owner and operator frustration—of evermore versatile and sophisticated control systems could easily outweigh their intended energy and labor cost saving benefits.

All controls do not represent state-of-the-art technology, however, and due to the long periods over which buildings remain in use, wide variation exists in equipment currently in use and being manufactured. Similarly, most commercial buildings now standing are equipped with traditional electromechanical and pneumatic controls, rather than the newer computer-based electronic systems.

As part of their sales efforts, controls manufacturers often provide a building energy analysis, if requested. The analysis, in the instance of new construction, provides the energy cost differences between various HVAC equipment and building operation alternatives. For existing buildings, the analysis determines the differences between pre- and post-retrofit energy costs.

Building energy analyses vary considerably, ranging from a complete audit to a simple survey of energy using equipment for existing buildings and, for new construction, from rough calculations to complex computer simulations. Often, the building owner has the greatest impact on determining how comprehensive or cursory the analysis will be. Owners and operators are often not willing to bear the expense of a detailed building energy analysis. Such a lack of adequate funding sometimes results in a controls manufacturer estimating post-installation cost savings to be greater or less than those that can be attained on an operational basis. The capabilities of manufacturers to provide accurate and comprehensive energy analyses varies from one firm to another, so that even those owners and operators who do pay for a full audit or analysis may not receive the quality of information they desire.

In addition to providing the controls system, the manufacturer or contractor also installs the equipment, and remains active on the project through an initial shakedown period. Manufacturers usually also train the building operator's staff in system operation and maintenance. The length and quality of both system debugging and building staff training varies substantially from contract to contract and from manufacturer to manufacturer.

As controls systems have become increasingly complex and sophisticated, the question of operator training has assumed major proportions in the field. As with building energy analyses, owners and operators are generally unwilling to invest in adequate and continuing training for their operating and maintenance (O&M) staffs.

There is no official source for the industry's total annual sales. One available estimate for 1979 indicates a total volume on the order of $1.16 billion, which includes both traditional and computer-based controls equipment. This figure would of course be considerably higher today.

The Department of Defense exerts considerable influence on the EMCS segment of the controls market. Involved in the purchase, installation and operation of EMCS since 1975, when $2 million was approved for six Navy locations, the Department has taken the lead within the Federal Government by developing a Tri-Service Guide Specification. The name has since been broadened to Interagency Guide Specification, in a move to make it the standard for all Federal EMCS procurements. The intent of the Navy, as the lead agency within DOD for EMCS, was to develop specifications that were ahead of the industry's position in terms of technology and service.

As with any initial effort in a new area, the first Guide Specification has been refined and modified. Many people active in controls resent the intrusion of the Government, and indicate that there are many problems remaining to be resolved with the Guide Specification and its application. Others see this effort as the first serious attempt to bring a measure of standardization to the field, pointing out that continued refinement, strengthening and use of the Guide Specification throughout the Government could eventually lead to reduced reluctance on the part of potential private sector controls purchasers to invest in new technology.

The potential market for controls is truly huge: every commercial and residential building could be in the market. Most buildings will eventually be able to benefit from some form of energy management system.

Estimates of market growth cover a generous range, from 15% to 30% with the rate of growth for large buildings and timeshared systems tending toward the lower end, while that for programmable controllers and systems for medium-sized and small buildings representing the upper range. The demand for conservation services is substantial. A study found that energy conservation expenditures for schools, hospitals, and other institutional buildings is expected to total $4.75 billion during the 1980s, with the EMCS segment of the controls industry accounting for $500 million alone. The Federal Government will continue to play a large role in the market; more than $500 million was appropriated for EMCS installations at military facilities during the 1970s.

CURRENT ISSUES

Education and Training

Perhaps the most dominant concern in the controls field at this time is that pertaining to education and training. There is a real and possibly widening knowledge gap between controls manufacturers and those who design (i.e., consulting engineers), install, purchase, operate and maintain controls systems. The full national potential for efficient energy end use in buildings is likely to go unrealized if this separation is not appreciably narrowed.

There are many consulting engineers who do not know enough about building controls to design an adequate energy efficient control system. This situation has its roots in formal engineering education. The mechanical engineering curricula of most universities include a course on system dynamics or automatic control theory. These courses do not deal specifically with the equipment and techniques used in buildings (e.g., programmable controllers, sensors, EMCS). Courses covering the function and application of microprocessors are available as elective or graduate courses at only a few universities.

Another major element in the knowledge gap concerns commercial building owners and operating managers (as distinct from operating engineers). Few building owners or operators are sufficiently knowledgeable in the controls area to be able to select design engineering consultants capable of providing adequate control system design documents. This situation is due, in part, to both the scarcity of competent EMCS design engineers and to the lack of available nonproprietary information concerning computer-based control systems.

Most building owners and managers have also yet to develop a full appreciation that the systems which control their HVAC and lighting equipment have increased in complexity by an order of magnitude. Operating and maintaining computers and computer-directed equipment often requires skills not yet developed by most building operating staffs. This problem has at least two aspects, one economic and the other psychological.

Building owners and operators have been quick to see the benefits that could accrue through the use of more sophisticated control systems. Yet, they cannot at present afford to hire graduate engineers to operate these new systems, even in the most complex new office buildings. For the most part, commercial buildings are still being run by operating engineers who belong to local labor unions. Similarly, owners and operators do not take full advantage of the courses offered by controls manufacturers. Instead of investing in adequate and continuing training for their operating staffs, they generally settle for the training provided by the manufacturer during the system installation and shakedown period. Additional training is required, however, for operating staffs to realize energy cost savings available and take advantage of the control and information benefits through optimal use of their control systems.

The psychological aspect of the situation stems from the general feeling that computers are basically labor-saving (i.e., job-threatening) devices. In many instances, an EMCS is an operating engineer's first exposure to computer-based technology in a daily, work-related environment. When an operating engineer joins a building staff at some point after the usual one-time training session provided by the controls manufacturer, he is likely to receive only whatever on-the-job training other staff members are willing to supply.

For those operating staff members who do participate in the manufacturer's training program, complex systems may be perceived as eventual labor-eliminating devices if the instructor assumes too much prior knowledge on the part of the building staff, or if the material presented is covered too rapidly. The regrettable current situation is that operating engineers sometimes allow control system computer programs to "crash," without taking corrective action, because they view computer failure as a form of job security.

Locals of the International Union of Operating Engineers (AFL–CIO) also play a role in their members' skill levels. In one major urban area, two union locals exist in adjacent states. One local, whose members already have the benefit of job experiences ranging from commercial buildings to factories to chemical plants and refineries, allocates more than \$200/member/yr. for education activities. The other union has only recently begun an educational program, spending about \$30/member/yr., despite the fact that its members' experience is primarily limited to commercial buildings. Situations like the latter contribute to a continuance of building operators who lack the requisite skills to run their buildings in an efficient and cost-effective manner.

Some professional societies and engineering consultants offer a variety of seminars, workshops and mini-courses in controls. There is a need, however, to reach larger audiences and to improve the quality of some of these offerings. Particularly needed are educational efforts targeted for building operators and building maintenance and operating staff members.

The National Electrical Contractors Association (NECA) has indicated that adequate training in controls is of concern to electrical workers and contractors alike. Members of the International Brotherhood of Electrical Workers are not, in general, trained to deal with electronics, black boxes and EMCS. The union is hesitant to train its membership in these areas unless it can be assured that there will be sufficient work to justify that training. At the same time, NECA members are equally reluctant to take on jobs in these areas unless they know there is an already trained labor pool to be tapped.

A concern touching graduate engineers and building operators alike is that of motivation. In addition to educational sessions, standard solutions to typical control system problems appear in technical and trade publications at regular intervals. Despite all these efforts, simple problems (e.g., beer cans jamming outside air dampers open) are routinely found during building energy surveys. How to motivate those who work in controls but who will not take advantage of available training and published information, is one of the most perplexing and important open questions at present.

In the same vein, building managers will not achieve their energy cost savings goals if they are unable to motivate their operating staffs to work with, rather than against or in indifference to, building control systems. More initial, and continuing refresher, training constitutes one means to accomplish this motivation. Controls manufacturers and building managers together have the responsibility to develop the perspective in operating staffs that they are an integral part of the building operation team, and that they will not be replaced by sophisticated energy management systems.

Another area in which education is needed concerns upgrading already installed equipment. Numerous problems exist in many commercial buildings that have poor maintenance programs. Maintenance staff turnover and lack of operational knowledge of their equipment have led to the attitude: "If it's working, let it alone; we don't want to disturb it." At different levels, consulting engineers, building owners, managers and maintenance staffs need to be made aware of the benefits (e.g., cost savings, fewer problems) that can be realized through upgrading the design, installation, operation and maintenance of existing controls systems.

User Satisfaction, Cost Savings, and Other Benefits

There is a clear relationship between some aspects of the controls knowledge gap and users' satisfaction with their installed systems. Complaints voiced by controls users can often be traced to education and training lacks in one or more areas: system design, operation and/or maintenance; preexisting controls problems; and/or building energy analysis. Energy cost savings are unlikely to reach anticipated levels if such deficiencies exist and persist.

Advances in technology through direct digital control have made it possible to achieve solid energy cost reductions while providing occupant comfort. However, sophisticated controls do not rectify deficiencies in the controlled systems. The trend in the controls industry has been towards maximizing both comfort and savings. To be successful in this approach it is imperative to address each of the points discussed under this heading.

User satisfaction also means different things to different people and, perhaps, even to the same person at other points in time. It has been noted that, although some commercial building owners and operators with EMCS experience have been quick to speak out on their systems' shortcomings, many of the same people quite readily admit that they are planning to purchase another EMCS in the foreseeable future.

The latter position was attributed to rapidly rising energy prices in the past. Cost saving is still a major reason for EMCS buying decisions. In spite of the downturn in oil costs, electricity has been affected only in the area of fuel charges. Therefore users, especially those with successful track records, continue to install systems. The continued success of automations systems goes beyond cost reduction. Buying decisions today are being influenced more and more by other system benefits. Communications and Information Management to aid in equipment performance and service, and non-energy related control features such as equipment protection, are among the key user needs.

The ultimate causes of user dissatisfaction with controls systems do not lie solely with one group. They can be traced to a number of groups, including designers, owners, and operators, as well as the oft-maligned industry itself.

User satisfaction, in short, is not likely to be without its very real costs. For example, greater component reliability is easily possible if a building owner or operator is willing to pay the higher price of industrial grade controllers. There is an obvious conflict between viewpoints. While engineers are interested primarily in increased reliability, building owners and operators are generally more concerned with rapid payback on capital investments. In building controls these interests often lead to different equipment selections and system cost estimates. Where the need for rapid payback prevails, system reliability may be an early casualty.

In spite of extensive system experience, building owners and operators still appear to believe that new controls systems, and EMCS in particular, are "miracle cures" for existing problems. When this attitude prevails, unmet energy saving and performance goals may follow new equipment installation. The reasons for such deficiencies can often be traced to previously existing system problems which have gone uncorrected, due to lack of maintenance and/or management attention.

Typically, however, the blame for poor system performance is attributed to the new equipment. As with so many of the foregoing areas, building owners and operators need to become more aware of their properties as integrated, and increasingly often as sophisticated, systems, which require both routine corrective and periodic preventive maintenance if they are to function at optimal performance levels.

The foregoing is not to suggest that the controls industry is without fault in the question of user satisfaction. While the established major manufacturers think in terms of controls and building systems, many of the newer entries in the field are product oriented. A successful controls system is one which integrates separate components (i.e., products) into building systems. Without this basic focus, individual elements sometimes work against each other, rather than in concert (e.g., by bringing in outside air on a 10°F day while the heating system is running at near capacity).

Also, the lack of acceptance tests to verify that installed systems meet design specifications and/or manufacturer's energy use claims can feed subsequent user dissatisfaction. The availability and consistent general use of objective, accurate and easily repeatable tests would go a long way toward reducing complaints about system performance.

The consistent experience of consultants investigating existing building controls is that systems which do not work as designed are poorly maintained, seldom calibrated, and often disconnected. It is therefore critical to develop an effective program management function as discussed in Chapter 17 with any automation program.

Controls Standardization

Building controls standardization has developed as one of the most pressing issues in the field today. The issue has been a point of contention for several years, and it is only recently that the industry has begun to address the critical points. Though both sides of the major disagreements are well supported, there is no clear-cut solution in sight that will completely satisfy all participants.

The primary issue stems from the need for users to support multiple front ends (programming and communications hardware and software) in order to utilize different manufacturers' equipment. Two schools of thought exist in relation to solving this dilemma. First, there are users who believe that at the most basic level all manufacturers' equipment should transmit system data and instructions using the same protocol. The difficulties associated with this scenario lie in the obvious requirement for manufacturers to share proprietary information. The second and more feasible option is for all automation systems to communicate on a universal protocol. This would allow manufacturers to perform the internal data operations in their own way, while offering users the needed communications flexibility.

One position widely held by manufacturers is that standards raise issues such as service and warranty responsibility, and jeopardize their competitive opportunities in the marketplace. This issue is covered in detail in Chapter 11.

The consensus industrywide favors standardization, for valid reasons. The existing lack of controls hardware and software (i.e. communications protocol) commonality often inextricably links building owners and operators who want to expand their systems, to the original equipment manufacturers, more often to the latter's advantage than to the former's.

At the same time, while it is possible in some instances for controls manufactured by different firms to communicate with each other, such situations are the exception rather than the rule. Even then, special software programming or additional interfacing equipment is usually required. These circumstances have led to some wariness within user communities, which are sufficiently aware of EMCS-related problems to be reluctant to depend entirely on any single manufacturer.

While standardization carries some problems of possible redesign to accommodate compatible hardware interfacing, the question of communications commonality presents a somewhat greater challenge. For protocols, standards will be needed which permit the capability to communicate without divulging proprietary information to other manufacturers.

The pro-standardization view has been established, but the method to be employed is still in question. Building owners and operators appear to require standards to assist them in making appropriate equipment selections for individual applications, and to lessen their reliance on consultants who may not be adequately familiar with their needs, or on manufacturers whose interests may be divided between corporate and customer demands. ASHRAE has taken the industry lead to develop standards with the convening of a recent Standards Committee. Representatives of each segment of the industry will be appointed to the committee to study the issue, and make recommendations for a course of action. Automation users and manufacturers alike will watch the results of this process carefully.

Standardization needs also extend to the areas of energy savings calculations and acceptance tests for installed equipment. If energy cost savings figures are to attain a degree of credibility with controls users, the accuracy of the calculations involved must be improved. One way to achieve this is to develop standardization calculations that can be performed by building owners and operators as well as by consulting engineers and controls manufacturers. Similarly, the lack of standard acceptance tests has made it difficult to verify that the performance of installed controls systems meets purchase contract specifications.

Design Integration

Various aspects of this issue have already been suggested. They are sufficiently important to be reiterated. Building design has traditionally been, and continues largely to be, a fragmented process. Typically, the architect designs the building envelope, then the mechanical engineers design the mechanical systems, and only then does the controls system designer's participation begin. By that point, it is exceedingly difficult to exert a major influence on energy use in the building, as most of the critical energy impact decisions will have already been made.

In those new buildings where design efforts to increase energy efficiency over the norm have been successful, an integrated design team has usually taken a major role. A fully integrated team, to take maximum advantage of all necessary inputs, should include the architect, the mechanical equipment engineers, the controls system designer, the electrical/illumination engineer, the building operator and, possibly, a representative of the major building occupant.

Illumination is a primary energy use in many commercial buildings, and early integration into the controls system will allow greater flexibility in building operation from a lighting standpoint. The building operator should also be called on during the first stages of design to provide experience-based input concerning operating procedures, maintenance requirements and energy conservation opportunities. Finally, if a primary building occupant has been identified, interaction with the design team will assure that the user's needs have been fully incorporated, and that the user is well aware of the building's potential for energy efficient orientation and concomitant energy cost savings.

EMERGING ISSUES

There are a number of questions that are developing into major topics of discussion within the controls communities.

Trend to Electronic Local Loop Controls

Pneumatic and electromechanical controls are gradually being replaced by electronic local loop controls. This trend will simplify interfaces with energy management systems. At the same time, it implies that changes will be required in traditional controls installation practice, and some retraining of building operating and maintenance personnel will be necessary. Controls systems are often installed by mechanical and electrical contractors. At least in the electrical area, most union members have not been trained in electronic equipment installation, and the union and NECA hold differing views of the proper timing of training in this work, as previously noted.

Increasing Complexity of Control Strategies

Integration of HVAC equipment, controls and building envelopes to achieve maximum fuel economy, and use of renewable energy resources (i.e., heat recovery, storage, solar collectors, wind, etc.) will require the use of controls strategies more complex than those in use today. Once again, this will increase present concerns over the differences in the quantity and quality of technical understanding of controls that exists between users and manufacturers.

Market Absorption of Innovation

The emerging issues just mentioned force consideration of a larger, more fundamental concern: that of the market's ability to absorb innovation in building controls. The importance placed on the question of adequate education and training for today's market, much less for the market of 1990 or 1995, indicates that a gap already exists.

As the rate of controls innovation appears likely to continue to outstrip the market's ability to absorb new controls concepts, this gap can be expected to widen further before efforts to narrow it have appreciable effect. And there are, at present, few serious efforts underway to bridge this rift.

Utility and Communitywide
Energy Monitoring and Control

While the industry has successfully opened up a new market segment of small commercial buildings, developments are also under way that could lead, at the other end of the scale, to an expansion of time-shared EMCS to coordinated, large-scale energy management.

A few innovative utilities are developing energy cooperatives, each of which includes several organizations with commitments to reducing their energy costs, and with varying energy demand patterns. An energy information center for each cooperative communicates with both the utility and each member organization via dedicated telephone lines. A databank at the information center has each member's energy use pattern recorded in a computer, thereby providing the knowledge of where it is the most feasible to reduce power demand when necessary.

When the utility's overall demand nears peak power production, members are instructed that certain kilowatt use reductions are required of them. Member operations managers determine where cutbacks can be made, and shed loads accordingly. Energy is saved by dimming lights, turning off escalators, alternating chillers in air conditioning systems or reducing whatever power is most expendable at the time. If a member is unable to satisfy the cooperative's power reduction requests, other members are asked to temporarily share the burden by consuming even less power. Initial results indicate that the cooperatives are manageable and effective in reducing energy demand.

Communitywide energy management is another approach to lowering energy use, this time in public buildings, by using the excess cable capacity of local cablevision companies. A community EMCS is linked to schools, libraries and other public buildings by cable channels, and energy use, especially during unoccupied hours, is monitored and controlled from a central facility. This type of effort is still in its infancy, but is expected by some to rapidly become very popular.

2

10 Reasons Why Some Automation (EMC) Systems Do Not Meet Their Performance Goals

1. Needs Not Defined

Although major EMC systems involve substantial capital investment, some have not been based on effective program plans. Little attention was given to actual needs, including those projected for the future which affect concerns relating to the expandability of systems and functions which they can be made capable of performing. Some EMC applications were not integrated into overall building energy management programs, to minimize duplicated effort and capital expenditures, while maximizing energy conservation.

2. Technology Not Understood

The technology used in today's EMC systems is still unfamiliar to many individuals who are called upon to develop specifications and who are responsible for their operation. As such, many operating engineers have been unable to: specify equipment with assurance that it best meets their needs; oversee installation; determine if the equipment is working to its potential; provide corrective maintenance beyond what is called for specifically in instructional manuals. Outside professionals who could provide needed guidance were not called in.

3. Lack of Standardization

Terminology and equipment operating techniques associated with new EMC systems vary from manufacturer to manufacturer. Multiplicity of terminology and techniques makes the technology confusing, especially for those who do not fully understand the systems involved. As a result, some operating engineers have been reluctant to call for proposals from numerous sources. When installing new systems and expanding existing equipment, they were not in a position to ask questions whose answers would help evaluate manufacturers' claims. In fact, input from various manufacturers made decisions far more difficult for them.

4. Manufacturer's Representatives Relied On Too Extensively

Operating engineers historically have relied on manufacturer's representatives to obtain information needed to make purchasing decisions. This can be an acceptable practice when conventional systems are involved, providing the operating engineers understand these systems, know what questions to ask, and contact technically-oriented representatives of several different manufacturers.

When it comes to EMC systems, the historical method often falls short of good practice.

For reasons already stated, operating engineers were often not in a position to compare systems or to ask questions needed for comparative purposes. They utilized proprietary specifications provided by manufacturers, which may or may not be suited to the specific facility involved. There may have been a need to rely on competent professionals for guidance—and none was obtained for that purpose. As a result, some installed systems have failed to meet expectations.

5. Systems' Potential Not Realized

The main benefit and economic justification of a computerized EMC system is systems optimization. However, many computerized EMC systems installed do not perform an optimizing function. This situation results from lack of knowledge and experience on the part of operating engineers and manufacturers representatives, and failure to use professional guidance.

6. Maintenance: A Problem

A central computerized EMC system must be kept in good operating order. The size of the market, newness of equipment, and sophisticated technology have so far limited the development of maintenance companies specializing in central computerized EMC systems. This can be a drawback of relying on a hybrid system. In addition, manufacturers of packaged systems warn that maintenance performed by other than manufacturer-authorized personnel may void their warranty or guarantee.

The high cost of maintenance is still another problem. The annual cost of manufacturer-supplied maintenance varies, but can range from 2 to 10 percent of full installed cost. Some operating engineers learn of these costs only after the system had been installed, indicating a failure to obtain accurate information about life-cycle costs.

The maintenance function for an automation system is also new to many users, because it entails much more than system hardware support. This function requires ongoing attention to the entire system including peripheral equipment. Unfortunately many users neglect system maintenance altogether, assuming that a computer system does not require this activity.

7. Absence Of Program Management

As with the maintenance of the hardware system components, it is also necessary to attend to the software facets of the EMC. Industry experience has confirmed that programs must be developed to verify system operation for ongoing system success. This function is critical to verify that the system is in control of the equipment, but can also contribute to savings through fine tuning. Users who are not in constant contact with the EMC are not likely to achieve system performance goals. This topic is addressed in detail in Chapter 17.

8. Training Inadequate

Those responsible for operating computerized control systems have, in some cases, not received adequate training. What has been provided relates almost exclusively to hardware descriptions and control system operation. As a result, in-house personnel are sometimes unable to use the computer to obtain a larger share of the potential which they offer.

9. Downtime: Excessive

Some systems have experienced frequent and extensive downtime problems. In some cases, responsiveness to calls for assistance has been good; in other cases, poor. Considerable time has been required to effect repairs by manufacturer-provided maintenance personnel who, some operating engineers have said, appeared to be "learning on the job."

10. Claims Not Substantiated

Occasionally, manufacturers' claims as to the energy-efficiency created by EMC systems cannot be substantiated. Although energy savings were verified, they could be attributed in large measure to modification of existing systems and their operation, as opposed to installation of EMC systems. The nature of the systems installed indicates that potential savings are not being realized due to lack of optimization functions.

The following chapters in this book present guidelines relating not just to the hardware and software associated with EMC systems, but also to steps that should be taken by user personnel prior to specifying a system, and even prior to making a determination that an EMC system is needed. Considerable discussion also is provided with regard to the way in which EMC and related systems work. This is done with purpose since lack of this understanding has in some instances led to specification of equipment not as suitable as alternatives available at the time.

3

The First Step to a Successful EMC System: Develop an Initial Energy Management Plan

A planned management approach is needed to govern the entire energy management function, which itself is a multifaceted process whose various activities are interrelated. The activities which comprise this planned management approach are shown in Figure 3-1, and are elaborated on through subsequent chapters in this book.

The planned management approach can be separated into two general work phases. The first phase relates to development and partial implementation of the initial energy management plan.

The initial energy management plan is developed on the basis of information derived from a comprehensive energy audit. It is the purpose of this plan to identify all ECO's and the order in which they should be implemented, including the conduct of feasibility analysis which may be needed.

Included in the category of ECO's are general items called AO's or Automation Opportunities. Securing a qualified automation consultant will assure that the user is informed of the many opportunities in this category. AO's must also be defined in the initial plan.

An AO is defined as a non-energy related system function which provides a benefit in the form of cost reduction, information, enhanced facility control or equipment protection. An example of an AO is outlined in Chapter 23 which details the integration of EMC and security systems for improved facility protection and control.

The initial plan should detail procedures which can be taken to eliminate energy waste and help optimize the performance of existing systems. In some cases, optimization can be achieved only through modification of controls and/or use of new control devices. In these

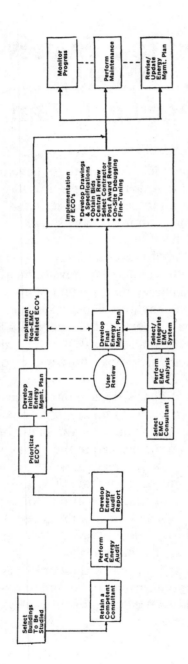

Fig. 3-1. Planned Management Approach for Energy Use Optimization in Existing Buildings

cases, details should relate to projected energy and energy cost savings, impact on other systems, and similar factors, exclusive of the specific control methodologies which may be applied.

In-house personnel should begin implementing the initial plan before the final plan is developed. This will help ensure that those involved obtain more understanding of energy conservation measures, especially so because new operating and maintenance routines are likely to be involved. In fact, improved operation and maintenance alone can result in a 15 percent energy savings.

The second phase of the planned management approach relates to development and implementation of the final energy management plan.

The final energy management plan is one which details all applicable ECO's, including those which relate to EMC modification and additions. The plan indicates what each ECO involves, when it will be implemented, projected costs and savings, and similar information.

As part of this phase, as well as the first phase, progress is monitored and the plan is updated in accord with experience.

By separating development of the final plan into two stages, more alternatives are identified, helping to ensure that the decisions made lead to substantial energy savings at the least capital expense, without impairing the facility's ability to accomplish its mission.

THE INITIAL ENERGY MANAGEMENT PLAN

There are many different types of EMC systems available. Some are comprehensive and integrate virtually every energy-consuming device in a facility. Others are local controls that operate on an independent basis. Deciding which particular type of system or systems is best for a given building is a procedure that can be effected well only through comprehensive familiarity with the equipment and systems which exist, the purposes which they serve, likely future changes (including expansion), and related concerns.

Of particular importance—no new EMC system should be installed until a plan for optimization of all existing systems has been developed. Then and only then will those responsible be in a position to determine which specific EMC systems are most applicable; then and only then will the systems designed, specified and installed be most likely to help the facility reduce its energy consumption and operating expense to the lowest level possible.

SELECT BUILDINGS TO BE STUDIED

The full range of ECO's present in a given building can best be identified by a competent, experienced consultant, working in conjunction with the operating staff. Before a consultant can be retained, however, the scope of work must be defined. In the case of facilities which comprise more than one building, scope of work must indicate specifically which buildings are to be studied.

Establish Building Selection Criteria

Appropriate facility personnel should work together to establish criteria to select buildings for detailed analysis.

As a general rule, those included should either

1. represent 5 percent or more of the gross square footage of all buildings, or

2. account for more than 5 percent of the total energy consumed for all buildings.

Additional criteria should be considered, as follows.

Energy Use Index (EUI): The energy use index, measured in Btu/ ft^2, indicates the relative energy efficiency of a building, when compared to others in the same complex. Those which are least efficient may be the ones most suitable for energy use optimization. Nonetheless, EUI factors should be used only as one of several criteria to rank buildings for detailed energy studies.

Substantial Total Energy Consumption: Substantial total energy consumption (more than 20×10^9 Btu/year) often indicates a significant savings potential, even with only a modest percentage of improvement.

Wide Range Between Maximum and Average Demands: A wide range between maximum and average demands, which cannot be accounted for by factors such as seasonal or day/night changes, may indicate a potential for more efficient use of energy and equipment. The range is identified easily as the ratio of monthly energy consumption to monthly peak demand, for example, kWh/kW for electricity. The optimum monthly ratio for continuous (24-hour) use is 720. The point usually indicative of potential savings is 400 or less. Proportionate adjustments should be made for shorter usage periods.

Similarity with Other Buildings: Some buildings may be identical to others, or have identical features. Substantial energy savings may be possible by applying techniques used for one building to other buildings of the same or similar type. To select the one building which will represent the others of a group of similar buildings, the consultant should consider the number of typical features, data available, individual metering, and related factors. Buildings which are similar in configuration but different in function may not be amenable to the same energy-saving techniques.

Recognizable Options: The applicability of low cost, easily recognized energy conservation options are obvious reasons to select a building for detailed study. Sometimes smaller buildings are high energy wasters and so may offer better opportunities for saving than buildings which consume large amounts of energy efficiently.

Previous Energy Studies: The results of previous energy studies of individual buildings or categories of buildings or systems should be reviewed.

Restrictions: Access to, or the ability to make modifications within some buildings may be restricted due to the critical nature of the work being conducted in them. These limitations may preclude a successful in-depth energy study and/or implementation options. Such buildings should be the subject of specialized energy studies performed directly by, or in close cooperation with local personnel.

Similarity of Energized Systems: A number of energized systems are likely to be common to a number of buildings. Examples are double duct or terminal reheat air conditioning systems, steam heating, once-through ventilation. Each of these systems may present particular energy conservation options. To take advantage of the leveraging effect of invested study time to energy savings potential, buildings selected for study should include as many typical systems as possible.

Preliminary Selection of Buildings: Based on general information gathered through review of documents, buildings can be grouped by function, type of construction, and similar major energy factors. The buildings can then be ranked according to total floor area. Small, one-of-a-kind buildings consuming less than 5×10^9 Btu/year may be eliminated from immediate consideration. Other one-of-a-kind, special function, restricted access buildings should be deferred for separate consideration. The physically largest and/or largest energy consuming, most representative buildings then can be identified for further study.

RETAINING A CONSULTANT

By and large, facility plant management personnel all are highly qualified individuals who are dedicated to performing their tasks well. The nature of energy management is such, however, that the degree of expertise and talent required to perform many of the start-up tasks often is unavailable in-house. In such cases, a firm generally will have to retain a consultant to help ensure overall familiarity with the three general systems which comprise any given facility. These three systems are:

Energized systems, meaning those which consume energy directly. These include HVAC (heating, ventilation, and cooling), lighting, SHW (service hot water), etc. Many of these systems can be modified substantially to reduce the energy they consume without affecting system output except, in some cases, to improve it.

Nonenergized systems, meaning those which do not consume energy, but which do affect how much energy is consumed. These include building envelope (composition and condition), roof glazing, etc.

Human systems, meaning those groups of people who directly or indirectly affect energy consumption. Such groups include facility management, operating personnel, maintenance personnel, tenants or other occupants, etc.

It is important to recognize that the consultant should not only be familiar with these three systems and their subsystems and components, but also with the interrelationships involved, since systems and their components depend on and react with other systems and components. As such, the consultant must have the ability to provide an overview, identifying the overall impact of any given change, including the way in which a change affects any and all other systems. This overview is particularly important insofar as maintenance of the building's ability to meet its function is concerned.

Selecting a Consultant

The procedure used for consultant selection should emphasize quality concerns, because the qualifications of consultants are of considerable importance to the overall success of the energy management function. Identification of competence and discussion of fee should be wholly separate functions. Those responsible may choose to invite selected firms or individuals to submit proposals.

Before any modification is approved, its applicability must be reviewed thoroughly in terms of the function of the item involved, how the modification may affect that function, and how that effect may somehow have negative impact on the building's ability to function. For example, environmental conditions provided in some areas such as computer rooms are critically more important than those provided in office areas, food service areas, lounges, etc. While energy waste should be eliminated wherever it occurs, relatively high energy consumption when necessary to support the function of critical areas is not a waste.

It also must be recognized that buildings must comply with numerous codes and other regulations. Those making recommendations must know exactly what codes are involved and the jurisdiction of each, and must ensure that any proposed modification is in keeping with applicable codes and/or the directives of code enforcement authorities.

Consultants may be found by contacting local chapters of national engineering organizations.

Consider the following suggestions for retaining the most qualified consultant:

1. Prepare a specification which identifies clearly and succinctly the work to be done.

2. Identify a group of consultants who seemingly are qualified.

3. Review initial submissions to identify those ten firms or so that have the type of experience required.

4. Select three firms and explain facility goals and objectives to them, allowing consultants to ask questions whose answers can provide them with the direction they need. Invite specific proposals.

5. After thorough review of proposals, and contact with clients for whom similar work has been performed by the consultant, rank the firms in order of preference, that firm deemed most qualified being listed first.

6. Open negotiations with the firm deemed most qualified. If discussions result in a fee proposal, scope of work and time-table which are mutually acceptable, the consultant should be retained. If negotiations are not successful, they should be closed with the top-ranked consultant, and opened with the consultant deemed second most qualified. The procedure should be continued until a competent consultant is found. If negotiations with the top three consultants are not successful, it could be that facility management may have to alter its fee limitations, scope of work, or other variables to help ensure that the consultant retained, in fact, is capable of performing the work needed in a timely, cost-effective manner.

Major contract concerns related to procurement of A/E services needed to develop a facility energy management plan, are shown in Figure 3-2.

This suggested contract is provided solely to indicate some of the major concerns which should be addressed. The contract ultimately used should be developed with your legal staff or counsel.

CONTRACT OBJECTIVE: To develop a plan for a facility-wide energy management program.

SCOPE OF STUDY: The study will include review of all building energized, nonenergized and human subsystems that significantly affect building energy consumption. These subsystems include, but are not limited to HVAC, electrical, lighting, domestic hot water, laundry, maintenance and operating procedures, building envelope, and control systems. The following buildings _____
_____ will be included in the study.

METHODOLOGY: The consultant will perform the following necessary tasks to satisfy study objectives.

TASK I: Meet with appropriate facility management personnel to review and finalize the scope of work, defining methodology to be followed and identifying a time frame for conduct of the study; development of the final report, submission of other materials, etc. All data will be provided to the consultant at this time, such data being energy use and cost records, as-built plans, specification, operating and maintenance manuals, etc. The consultant will identify additional information needed. Protocol for conduct of inspections and interviews with management, department heads, engineering and operating personnel, will be discussed.

A final schedule will be submitted after review.

TASK II: Conduct a comprehensive energy audit of the buildings identified in the scope of work. Energy audit must consider, as a minimum, the collection and review of data provided in Table 3-1. Additional information relative to review of energy conservation option is to be collected as required. The audit will result in identification and definition of energy conservation opportunities (ECO's) applicable by buildings and their respective subsystems.

TASK III: The consultant will analyze and categorize various energy conservation options using life-cycle costing techniques. ECO's shall be prioritized in terms of the plan.

TASK IV: Develop a preliminary report, typewritten, double spaced, and using plain and concise language which defines the study objectives; delineates the approach used by the consultant and provides analysis of the various ECO's being recommended. An executive summary is recommended.

Detailed calculations related to modifications will be provided separately, or in an Appendix.

TASK V: Incorporate owner comments and develop a final report.

SCHEDULE: The following schedule will be followed:
- Preliminary Meeting—10 working days after award of contract
- Energy Audit Completion—30 working days after award of contract
- Preliminary Report—60 working days after award of contract
- Final Report—90 working days after award of contract

DELIVERABLES: The following deliverables are required under this contract.
- Preliminary Report—10 copies
- Final Report—10 copies

Fig. 3-2. Major Contract Concerns for Procurement of A/E Services Needed to Develop an Energy Management Plan

CONDUCTING AN ENERGY AUDIT

"Energy audit" is the term used to identify procedures typically used to develop the report which identifies available ECO's. A list of the basic types of information required for a detailed energy audit is shown in Table 3-1. The list is far from complete, but does serve to indicate the different types of concerns which must be addressed.

Developing the information involves two distinct types of procedures: The review of documents (energy consumption data, operating logs, plans, etc.), and a physical inspection of facilities and equipment, including discussions with and observation of appropriate personnel.

The review of documentation usually occurs first, to provide an overview of what is involved in general.

Facility staff should provide the consultant with such information and materials as generally are required for the development of an effective energy management report. Of primary concern are energy use records, available from facility, utility or fuel suppliers' records. These records permit the consultant to develop an "energy use index," or EUI. The EUI measures energy consumption in terms of Btu/gross conditioned square foot/year (or month). The form used for this purpose is shown in Table 3-2. Other materials which are needed include original plans and specifications, as-built drawings (if they exist), operating and maintenance manuals and logs, and such other information and materials as the consultant may request.

Once a study of documents and related tasks has been completed, a physical inspection can be conducted. Through this procedure, consultants can identify the overall nature and specific elements of the three basic systems (energized, nonenergized and human), and the way in which they interact with one another to consume energy. In addition, the consultants can identify the condition of energized and nonenergized equipment, and such other factors which may be used to suggest the applicability of various ECO's (energy conservation options).

Table 3-1. Energy Audit Considerations

Identity and Use
    Name
    Use

Physical Data
    Floor area
    Number of stories
    Roof area
    Roof construction
    Wall construction
    Window type
    Glazing
    Shading employed
    Floor construction
    Exterior building dimensions

Operating Schedules
(weekdays, Saturdays, Sundays, and
holidays)
    People
    Lights
    HVAC
    Process
    Custodial
    Etc.

Energy Sources
    Electricity
    Gas
    Oil
    Steam
    Purchased chilled water

Energy Cost Data
    Electricity demand & consumption
    Gas
    Oil
    Steam
    Purchased chilled water
    Rate schedules

Historical Monthly Energy Demand,
Consumption Data for Past 2 or 3 Yrs.
    Electricity
    Gas
    Oil
    Steam
    Purchased chilled water
    Coal

Standby Power/Energy Sources
    Type
    Capacity

Electrical Characteristics
    Voltage
    Power factor

Type of Heating
    Space
    Domestic and service water

Type of Cooling
    Space, Process

HVAC System
    System number
    Area served
    Critical/noncritical
    Type of air-side system
    Type of water-side system
    Type of control and existing
        control device
    Outside air
        Minimum required
        Maximum available
    Measured running amps

Domestic and Service Water
    Size
    Rated input
    Aquastat setting
    Usage
    Heat recovery application

Other Energy Consuming Devices
    Item
    Energy demand
    Operating requirements
    Existing control
    Critical/noncritical functions
    Measured running amps

Pumps
    Service
    Capacity
    Critical/noncritical
    Measured running amps

Existing Storage Tanks
    Item
    Use
    Capacity
    Critical/noncritical

Other Existing Communications
Systems
    Existing capability
    Spare capacity

Telephone System
    Existing capability
    Spare capacity

Table 3-2. Energy Management Form

Building _____

Gross Conditioned
Square Feet _____

| MONTH 1 | HEATING DEGREE DAYS 2 | COOLING DEGREE DAYS 3 | PURCHASED ELECTRICITY | | | | | | | | |
			kWh COST 4	DEMAND CHARGE 5	POWER FACTOR ADJ. 6	FUEL ADJ. 7	TOTAL COST 8	BILLED DEMAND (kW) 9	ACTUAL DEMAND (kW) 10	kWh USED 11	BTU's 12
JAN.											
FEB.											
MAR.											
APR.											
MAY											
JUNE											
JULY											
AUG.											
SEPT.											
OCT.											
NOV.											
DEC.											
YEAR TOTAL											

Btu Conversion Factors: Electricity, kWh—3413; Purchased Steam, 10^3lb—1,000,000; Natural Gas, 10^3ft^3—1,030,000; Oil (No. 2) gallons—138,700; Oil (No. 6) gallons—149,700

Table 3-2. Continued

MONTH 1	PURCHASED STEAM							OIL, HEAVY			OIL, LIGHT		
	STEAM COMMODITY COST 13	STEAM DEMAND CHARGE 14	TOTAL STEAM COST 15	DEMAND ACTUAL (lbs/hr) 16	DEMAND BILLED (lbs/hr) 17	STEAM USED 10³ LBS 18	BTU's 19	TOTAL COST 20	GALLONS USED 21	BTU's 22	TOTAL COST 23	GALLONS USED 24	BTU's 25
JAN.													
FEB.													
MAR.													
APR.													
MAY													
JUNE													
JULY													
AUG.													
SEPT.													
OCT.													
NOV.													
DEC.													
YEAR TOTAL													

Table 3-2. Continued

MONTH 1	NATURAL GAS					COAL			TOTALS			
	COMMODITY COST 26	DEMAND CHARGE 27	TOTAL COST 28	10³ FT³ USED 29	BTU's 30	TOTAL COST 31	TONS USED 32	BTU's 33	TOTAL COST 34	COST FT² 35	TOTAL BTU's USED 36	EUI 37
JAN.												
FEB.												
MAR.												
APR.												
MAY												
JUNE												
JULY												
AUG.												
SEPT.												
OCT.												
NOV.												
DEC.												
YEAR TOTAL												

The participation and cooperation of facility operations personnel and of the buildings' management staff is critical to development of effective ECO's. Since in most cases limited hard data is available, soft data will have to be developed. As such, the energy study team will be largely dependent on data provided by building operations and maintenance staffs. During the physical aspect, the energy study team should be accompanied by building personnel to provide data on operation and maintenance and on the feasibility of the potential ECO's identified.

DETERMINE ENERGY SAVINGS

The method used to determine energy savings and the impact on other systems of any given change also is a matter of concern. Approximations should not be relied upon, except in the case of relatively minor changes. It is recommended, therefore, that you utilize either a degree day approach method, as discussed in publications of ASHRAE (American Society of Heating, Refrigerating, and Air-conditioning Engineers), or more complex computer simulation techniques, using one or more of the many different simulation programs now available. These programs are designed to simulate a building's energy systems using refined input such as hour-by-hour weather tapes. To evaluate the impact of an ECO, the program is first run without the ECO, and then with the ECO. The energy saved is the difference in energy requirements calculated in the two runs. The program can then be re-run with different combinations of interrelated ECO's, until the energy savings are maximized.

CONSULTANT'S REPORT

The consultant's report forms the basis for the initial energy management plan. A typical summary report format is shown in Table 3-3.

It must be recognized that the consultant's report will include ECO's which represent alternative approaches to modifying a given system, each approach having its own merits and cost considerations.

Table 3-3. Consultant's Typical Summary Report Format—Initial Energy Management Plan

Building	Option	Initial Costs ($)	Annual Savings (Btu)	Annual Savings ($)	Payback Period (Years)	Energy Source
No. 10	Preheat Combustion Air	21,000	$7,178 \times 10^6$	13,800	1.52	Oil
	Replace Worn Boiler Controls	8,750	$4,785 \times 10^6$	9,235	0.95	Oil
	Reduce Air Volume	23,000	$1,587 \times 10^6$	16,250	1.42	Electricity
	Install Automatic Thermostats	800	90×10^6	170	4.70	Oil
	Provide Lighting for Specific Tasks	8,000	254×10^6	3,200	2.50	Electricity
No. 15	Use More Efficient Fluorescent Lamps	54,000	$1,536 \times 10^6$	15,760	3.42	Electricity
	Reduce Air Volume	17,000	$1,024 \times 10^6$	10,500	1.62	Electricity
	Install Switching	32,000	614×10^6	6,300	5.08	Electricity

In these cases, the consultant should also make comments related to ECO's whose application will affect others, for example, modifying ventilation rates will affect exhaust system ECO's.

Cost considerations are important. The consultant must know beforehand which factors should be considered, and the specific measures which should be applied, such as simple payback, life-cycle cost, benefit/cost ratio, or others. These concerns are discussed in Chapter 15. The calculations used in deriving the figures shown, including those for energy savings and projected energy cost increases, should be included in a separate section of the report.

The report also should include general, conceptual schematic diagrams of certain major changes. The schematic would indicate approximately how the modification would be made, to ensure understanding of the ECO.

PRIORITIZE ECO'S

The consultant's report will list numerous ECO's. Some modifications will be amenable to rapid implementation by in-house personnel, at little or no cost. Others will require outside assistance and considerably more expense. Still others will have to be studied closely to determine which alternative is best and, once that decision is made, design and specification will follow. The process of evaluating options, and determining the sequence in which they will be implemented, is called prioritization.

The prioritization functions should be performed by a committee or team of individuals who can provide diverse types of input. The consultant should be on hand to provide expert guidance and answer questions.

The chief operating engineer should be on the committee because he will be responsible for implementing most of the ECO's, and must be concerned about scheduling requirements and other matters which affect the timing of work and the ability to carry on routine functions.

A representative of building management should be on hand to provide guidance particularly with regard to budgetary concerns, the amount of capital improvement funding available or to be sought, and other fiscal matters. In addition, management can provide guidance on physical changes which may be planned, whose implementation will affect ECO's. For example, the addition of new equipment may result in more demands being placed on an existing system, thereby suggesting that one change would be more appropriate than another.

Certain department heads should be involved as well, but they may not be needed for the entire prioritizing process. Involving department heads in the energy management program from the outset also helps ensure their cooperation later on. Representatives of major tenants should also be part of the committee.

It is suggested that ECO's first be analyzed in terms of their inter-relationships. As such, it may be shown that it will be far more beneficial to implement five air-handling system ECO's at the same time, because it will greatly reduce the expense otherwise involved.

It is suggested that the first ECO's to be implemented should be those which will result in relatively substantial savings with little or no capital investment, especially those relating to bringing equipment not slated for replacement up to design conditions, and implementing new operating and maintenance routines.

FINALIZE AND IMPLEMENT
THE INITIAL PLAN

Most of the initial plan will remain intact and will be incorporated into the final plan. Accordingly, the initial plan must be developed with care.

In most cases, it will be appropriate to have a five-year plan, with close detail being provided for all ECO's to be implemented for the first 18 months, with more general comments being provided for ECO's to be implemented in months 19–60. At the end of every six months, close detail would be added for the next six months and, if appropriate, the plan would be extended by another six months to keep it a five-year plan.

For each ECO to be implemented in the first 18 months, the plan should indicate:

1. A description of the ECO in general terms;

2. the equipment which will be affected directly by implementation of the ECO, the nature of the effects, and other systems, equipment or operation which will or may be affected due to system interrelationships;

3. the estimated cost of implementing the ECO in terms of equipment, materials, and labor;

4. the amount of energy savings projected;

5. the dollar savings projected;

6. timing/scheduling factors.

Scheduling factors are of extreme importance, because they relate to the deployment of personnel and funds. Unless schedules are prepared accurately, problems will result during the implementation phase.

Some of the ECO's will require little scheduling, and may be applicable on a time-available basis. Others may require such steps as:

1. Retain consultant to perform feasibility analysis;

2. decide upon the specific ECO in light of the feasibility analysis report;

3. obtain funding;

4. retain consultant to develop plans and specifications;

5. retain contractors;

6. perform installation.

Each of these steps would have to be factored into the plan. In some cases, a year or more may be involved.

One of the steps which must be considered in the initial plan is conduct of the EMC analysis which will affect most ECO's which relate to major control modifications. This will not affect certain major ECO's. If the savings to be derived from implementing such ECO's are substantial, some of them could be started upon immediately. Conversely, if a major modification may be affected by EMC systems, it would not be advisable to schedule them for implementation until after the EMC analysis is performed.

Although the results of the EMC analysis may change the start-date of a major modification, they will not appreciably affect scheduling concerns in terms of the amount of time which it will take to implement the ECO from start to finish.

Once the initial plan is complete, it should be submitted for approval and funding. In the interim, certain of those modifications requiring no capital investment can be pursued; those requiring funding approvals would be implemented only after the initial plan is approved.

It should be noted that the plan can go beyond implementing ECO's. For example, it could indicate the dates for energy management committee meetings; staff assignments with regard to collecting energy consumption and related data; programs for development of energy awareness among operating personnel and occupants; and such other non-ECO actions which affect the overall energy management effort.

A format for an energy management plan is shown in Table 3-4.

Monitoring

Monitoring progress is an essential element of energy management, so that projections can be compared to actual experience. The information needed consists of, among other things:

1. The amount of energy consumed, by type, on a monthly basis;
2. the modifications actually made and when they were made, by whom, and actual costs involved;
3. any reports of unsatisfactory system performance.

Energy consumption can be tracked through the use of the energy use index (EUI) which reports energy consumption in terms of Btus/ gross conditioned square-foot/year (or month). Comparing energy consumption from month to month, or from period to like period of a different year, will serve to provide an excellent indication of the effectiveness of energy management procedures. Note that some consideration must be given for climatic variables. In other words, if the winter of 1983 was half again as cold as the winter of 1982, energy consumption figures for the respective periods would have to be adjusted for purposes of an accurate comparison to indicate the effectiveness of modifications.

Table 3-4. Energy Management Plan

Location	Description	Estimated Cost	Estimated Annual Savings	Payback Period	Estimated Life of Installation	Estimated ROI	Date of Modification	Personnel Responsible
Recreation Area	Modify existing heating system and controls	$15,000	$9,000	1 yr., 8 mos.	12 yrs.	51.7%	5/17	D.H.
Dining Hall	Split Econostat zone and balance heating system	$ 4,300	$1,700	2 yrs., 6 mos.	12 yrs.	31.2%	6/4	W.J.
Research Facility	Modernize existing heating system and install automatic controls	$ 9,000	$1,700	5 yrs., 4 mos.	15 yrs.	12.2%	6/23	N.K.
Library	Service, repair or replace steam valves and traps	$12,000	$3,000	4 yrs.	5 yrs.	5.0%	7/12	J.B.

4

The Second Step: Develop and Implement the Final Energy Management Plan

Once the initial plan has been finalized and experience with it has been gained, it is appropriate to study the feasibility of application of EMC systems. This evolves into a two-part process.

FIRST PART

The first part of the process requires retention of a qualified EMC systems consultant to review the preliminary experience and related concerns to determine which types of systems will be most appropriate. Types of EMC systems then are identified, and the effect of their application is studied in terms of other ECO's already included in the initial plan. In some cases, implementation of EMC-related ECO's and AO's will change initial scheduling, or may even suggest different priorities. Experience gained with the initial plan, and understanding of the ECO's it involves, will make development of the final plan relatively uncomplicated.

Major contract concerns for procurement of A/E services related to development of the final energy management plan, integrating EMC system considerations, are detailed in Figure 4-1.

The second part of the process involves actual design and specification of the EMC systems. If a combination of several smaller systems is to be used rather than one larger system, and if they are to be installed over time, development of plans and specifications is

Note: This document is provided solely to indicate some of the major concerns which should be addressed. The contract ultimately used should be developed with your legal staff or counsel.

CONTRACT OBJECTIVES: To develop a plan for a facilitywide automation and energy management program, incorporating all appropriate elements of the initial plan and supplementing them with EMC system considerations.

SCOPE OF STUDY: The study will include:

- Review data gathered during development of the initial plan, updating such data as appropriate;

- review the existing long-term plan, energy conservation opportunities (ECO's) and automation opportunities (AO's) which have been made in accord with it;

- review and identify condition of existing systems discussed in the existing plan, in consideration of ECO's scheduled for subsequent application to such systems;

- identify EMC systems which are applicable in light of existing conditions and proposed changes and for each, develop financial and energy data;

- prepare a final plan which incorporates appropriate elements of the existing plan and approved plans for application of EMC systems.

METHODOLOGY: The consultant will perform the following tasks:

TASK I: Meet with appropriate facility operating and management personnel to review and finalize the scope of work; define methodology to be followed, and identify the time-frame for conduct of the study and receipt of deliverables, etc. All appropriate materials will be provided to the consultant at this time, such materials to include those developed for the initial energy management plan, energy consumption records for that period subsequent to those studied for purposes of development of the initial plan, records relating to ECO's implemented in accord with the initial plan, and such other appropriate materials as are appropriate to determine the feasibility of applying various candidate EMC systems. The consultant will identify such other information as may be required for the purpose.

Inspection requirements will be identified by the consultant, along with identification of individuals with whom interviews will be conducted. Protocol for conduct of inspections and interviews will be discussed.

A final schedule will be submitted after review.

TASK II: Review buildings and systems and identify functions which should be provided by EMC systems, the appropriate level of EMC system required, and the specific type of system within a given level(s).   *(continued)*

**Fig. 4-1. Major Contract Concerns for Procurement of A/E Services
to Develop an Energy Management Plan**

TASK III: Develop and provide appropriate contract documents:

Drawings

- Title Sheet
- Drawing List
- Legend and Symbol List
- Site Drawing—Buildings and Data Transmission Cable(s) Routing
- Data Transmission System Network Diagram
- Interconnecting Trunk Wiring Data Link Diagram
- Point Schedule
- Functional Layout
- System Schematic Diagram
- Typical Control Wiring Diagrams—Interfaces and Interlocks
- Typical Control Sequence Description
- Floor Plans—Each Building or System
  1. Sources of Available Electrical Power
  2. Location and Nameplate Data of Equipment to be Monitored & Controlled
  3. Location and Type of Existing Controls and Starters
  4. Location of FID's and Data Transmissions or System Terminations
  5. Location of Radio Equipment
  6. Related Mechanical/Electrical System Modification
- Layout Drawings
  1. Typical Underground Data Transmission Wiring Details
  2. Detailed Power Wiring from Available Electrical Sources
  3. Typical FID and MUX Installation Diagrams
  4. Typical Instrumentation Installation Wiring
  5. Typical Aerial Data Transmission Wiring
  6. Typical Radio Equipment Installation Details
  7. Data Transmission Wiring Building Entrance Methods

Specifications

- EMC System Specifications

TASK IV: Provide the following services prior to, during, and following construction:

- Prequalify prospective contractors, and be available to answer any questions they may have;

- evaluate proposals and assist in selection of contractors;

- assist in development of contract performance planning and communication methodologies;

- provide on-site visitation during installation, as needed;

(continued)

Fig. 4-1. Continued

- assist in development of the acceptance testing procedure, and provide as-needed visitation during acceptance testing;

- participate in on-site debugging and fine tuning procedures for a period of one year after installation is accepted;

- review manuals and training program outlines for adequacy; and

- participate in energy management committee meetings for a period of one year after installation of system is complete.

DELIVERABLES: Provide three copies of each of the deliverables identified below in accord with the general timetable indicated:

- FIRST SUBMITTAL: 30% Design.
  1. Provide preliminary engineering and economic analysis of EMC system application.
  2. Indicate which systems and functions will be considered for EMC system application.

- SECOND SUBMITTAL: 90% Design.
  1. Indicate location of FID's.
  2. Provide an HVAC mechanical diagram depicting the individual HVAC components being controlled, along with the relative location of the sensors and actuators.
  3. Provide an automatic temperature control diagram depicting the required EMC system inputs and outputs in addition to the local controls required for optimized operation of systems and equipment.
  4. Provide a complete sequence of operations for the local controls, including the EMC system inputs and outputs.
  5. Provide a marked specification for the EMC system.
  6. Provide final engineering and economic analysis of EMC system application.

- FINAL SUBMITTAL: Project Design.
  1. Provide any corrections, additions, or deletions noted on the second submission.
  2. Provide final drawings and specifications.

Fig. 4-1. Continued

called for on a "one-at-a-time basis," thereby enabling the consultant to specify the most advanced system in use at the time involved.

PERFORM AN EMC SYSTEMS ANALYSIS

Energy management and control (EMC) systems should be selected to optimize the performance only of systems which already have been made as efficient as possible through conventional means, or which are scheduled to be modified to achieve that end.

There is tremendous range in EMC systems, categorized below into three distinct levels of control. (Locally-applied control devices are not included.)

Most of the EMC systems and devices now available include electronic components. Since developments in the field of electronics are progressing at an unprecedented rate, many systems and devices introduced just a short while ago rapidly are becoming obsolete. This situation demands that those responsible for preparing recommendations relating to the installation of EMC systems have a thorough knowledge of what currently is available, as well as a given facility's specific needs, in light of energy management plans and functions already established, as well as other plans for facility renovation, modernization or expansion.

It is recommended strongly that a consultant be used to provide guidance relative to EMC systems. If the consultant used for development and preparation of the initial energy management plan is qualified, reliance on the consultant's experience is logical. The consultant already will be familiar with many of the concerns involved. If the consultant is not sufficiently qualified, however, another should be retained, using methods already suggested.

Manufacturer's representatives should not comprise the sole source of input for EMC systems analysis. Inputs from several can be very helpful in establishing the parameters for an EMC system.

Review EMC-Sensitive Factors

Determining which type of EMC device or system is best requires an understanding of those factors which bear on EMC equipment selection.

The review should consider planned changes. For example, if a chiller is to be modified, the report should indicate the current situation, the date at which modifications will be made, and the resulting changes at certain times.

In addition to knowing what the various energized systems are, the consultant also will need to know which of them are capable of being controlled. Some may not be suited to automatic control.

Existing controls themselves also must be defined and accounted for. These include controls for air handlers, boilers, chillers, environmental control and other process controls. The consultant must know not only how, say, the air conditioning is controlled, but also how processes are controlled.

If modulating outside-air/return-air dampers are already part of the system, it will be inexpensive to convert to a Level III Energy Management and Control system. If there is only a minimum or fixed outside-air system, however, it may be costly to add an outside-air cycle. Duct openings would have to be cut into the walls, existing ductwork may have to be rebuilt, powered exhaust and recirculating air systems may have to be added, and so on. A number of energy (or media) control possibilities exist: a modulating valve, an open-and-close valve, a three-way bypass, primary/secondary pumping loops, etc.

The existing controls audit also may show that the building already is equipped with compressed-air-actuated pneumatic controls, or voltage-actuated electric controls that cause control apparatus to modulate, or electronic controls that operate valves or dampers. The problem then, is integrating already existing control devices with a new EMC system. Major changes may be required for compatibility.

THREE LEVELS OF EMC SYSTEMS

Categorizing EMCS into three levels may seem to be an oversimplification. However, the term is not used here to refer solely to number of points, or system cost. Rather, "level" is used qualitatively, to describe the degree of system sophistication. Appendix C gives quantitative categories. Succeeding chapters will evaluate each of the three levels of EMC systems. Greatest emphasis will be placed on Level III systems—engineered control centers.

Level I: Remote Limited and Multifunction Controllers

Level I EMC systems are typified by a limited function demand or load controller which interfaces with numerous energy consuming devices and systems to provide control of functions and/or limit electrical demand. Many of the newest Level I controllers are programmable microprocessors, and provide for multifunction capability in the same "black box."

Typical Level I EMC system functions include:

1. TOD start/stop load control, turning systems and devices on at a certain time of day (TOD);
2. optimized start/stop load control (based on outdoor and indoor temperature variances);
3. supervisory temperature control (interrupting local thermostat or control loop based upon sensor feedback);
4. status monitoring (Is it running or has it stopped?);
5. status alarms (Is it operating outside of the normal range? Is it too high? Too low?);
6. demand control;
7. duty cycling; and
8. remote communications (may be optional).

Level I systems usually are most applicable when:

1. the capital budget is $5,000 to $25,000;
2. the functions to be performed are limited; and
3. the number of points to be monitored and controlled are generally less than 50.

Level II: Central Monitoring and Control System

Level II systems are rarely encountered today, as they were extremely limited, and most of these were retrofitted during the energy conscious 70's and early 80's. These systems were designed to monitor and control all or selected energized system components by a central processor. The central processor does not include a computer, but it does include a graphic display panel which enables various manual and some automatic functions. For example, it can be programmed to turn off all boilers when the temperature is 65F and then turn them on again when the temperature falls to 60F. Note however, that the central processor cannot be programmed to handle the many variables which a computer-based central processing unit can.

Typical Level II system functions include:

1. remote manual start/stop;

2. automatic start/stop;

3. status monitoring to indicate whether the connected device is energized or de-energized;

4. status alarms, to annunciate (by sound, lights, or both) an off-normal condition of temperature, pressure, humidity, etc.; and

5. demand control or load-shedding equipment, and load-cycling controls.

Level II systems usually are most applicable when:

1. more than 100 points are involved;

2. optimization functions are not required;

3. multiple buildings are involved;

4. relatively few operating and maintenance personnel are on staff;

5. monitoring and control are performed from a central location;

6. all automatic control functions are preprogrammed;

7. control decisions are based on monitoring single or just a few parameters and conditions; and

8. no major HVAC system changes are planned to improve their level of sophistication.

Level II systems may still exist in some buildings. Generally speaking, they represent a state of the art which has now been surpassed.

In some cases, however, it is possible to upgrade an existing Level II system to Level III, or to at least utilize some of their components to reduce the cost of upgrading to Level III.

Level III: Engineering Control Center

Level III systems are similar to Level II, except one or more computers are utilized to make control decisions based on operating data and programs, as well as data contained in memory.

Level III system functions include:

1. automatic start/stop;
2. closed-loop refrigeration and boiler system control;
3. status monitoring;
4. status alarms;
5. load shedding;
6. duty cycling;
7. data collection;
8. enthalpy control;
9. event initiation;
10. utilities monitoring;
11. maintenance management;
12. monitoring and control of the building's life safety and security systems;
13. optimization of system operation to minimize energy consumption; and
14. remote communications.

Level III systems are appropriate when:

1. a capital budget of $30,000 or more is available;
2. optimization functions are needed;
3. control decisions are to be based on the number of parameters and conditions involved, and on the series of events which occur;

4. operating and maintenance personnel who understand the system are on hand or will be hired. (This is not an essential requirement where a remote communications option is put to effective use.);

5. user programming capability for parameters, and with some systems control algorithms, is necessary;

6. the facility is remote to the end user, is greater than 30,000 square feet, comprises a high rise building or a complex of buildings with a central plant and several remote mechanical equipment rooms;

7. there are 75 or more points; and

8. remote communications for monitoring and programming is critical.

IMPLEMENT THE FINAL PLAN

The final plan is essentially similar to the initial plan, except all feasible ECO's—including EMC-related ECO's—and AO's are identified and a timetable for their implementation has been prepared.

The final plan is implemented much in the manner that the initial plan is. Of key importance are the functions of maintaining comprehensive and accurate records, monitoring progress, keeping building personnel up-to-date on progress to maintain their cooperation, updating the plan on a regular basis, etc. In fact, if the initial plan has been implemented properly, implementation of the final plan becomes nothing more than continuing in the same manner as before.

5

Level I EMC System: Multifunction Supervisory and Demand Controllers

Level I control devices are typified by: 1) Supervisory Load Controllers which provide simplified time of day (TOD) and control loop interruption functions, and 2) demand controllers which interface with numerous other devices and systems to limit electrical demand. These control devices are programmable due to the use of microprocessors, and are multifunction controls, providing functions of localized controls (e.g., remote start/stop, optimal start/stop, duty cycling, etc.) in the same "black box."

The high value of savings which a multifunction load or demand-based controller can generate, coupled to low cost, make this type of control one of the most beneficial available.

A specification for a Level I EMC system should include factors listed in Table 5-1.

The two basic types of systems will be discussed separately in this chapter.

SUPERVISORY LOAD CONTROLLER

There are numerous multifunction supervisory type load controllers available on the market today. They are microprocessor-based and typically perform the following functions: temperature-compensated supervisory control of HVAC, electrical demand control, load scheduling based on time of day, optimized start/stop and duty cycling.

Table 5-1. Level I EMC System Specification Factors

I. GENERAL REQUIREMENTS
1. Application
2. Scope of Work
3. General Description
4. Existing Controls
5. Contract Documents
6. Contract Performance
7. Work Performance Schedule
8. Engineering Drawings and Specifications
9. Input/Output (I/O) Summaries
10. Equipment and Materials
11. Installation
12. Submittals and Documentation
13. Operator Training
14. System Requirements
15. Work Hours
16. Interruption of Operation
17. Warranty Service
18. Guarantee
19. Maintenance

II. SYSTEM & EQUIPMENT DESCRIPTION
1. Central Processor Electronics— Microprocessor
2. Central Panel Features
   a. Data Display
   b. Status Indication
   c. Individual Load Data Entry
   d. Key Lock Entry System

3. System Inputs
   a. No. of Analog Inputs
   b. No. of Digital Inputs
   c. Spare Capacity
4. System Outputs
   a. No. of Digital Outputs
      1. Pulse Width Modulation
      2. Dry Contact
   b. Spare Capacity
5. Peripheral Device Interface
   a. Printer
6. Signal Transmission Media
7. Equipment Options
   a. Battery Backup

III. SYSTEM FUNCTIONS & SOFTWARE
1. Demand Control
2. Duty Cycling
3. Supervisory Temperature Control
4. Optimized Start/Stop
5. Morning Warm-Up or Cool-Down Outside Air Damper Control
6. Nighttime Damper Control
7. Enthalpy/Economizer Control
8. Water Temperature Reset
9. Freeze Alarm
10. Alarms (Temperature, Humidity, Velocity, Pressure)

Functionally, all comparable programmable controllers are basically the same. But the versions offered by various manufacturers are quite different.

Difference in versions that are available for off-the-shelf purchase lies in these basic areas: memory capacity, method of programming, input/output ratings, expansion capability, expansion increments, readout options and methods used to accomplish end results, and remote communications options.

These controllers are used for automatic scheduling of various electric heating and cooling units and other electrical loads in a building. In the past a typical controller package included an integrated circuit logic panel and two remote sensors, one for outside temperature, and the other for indoor temperature at night. Both sensors are connected to the logic panel by low-voltage wiring.

The programmer operates by switching a limited number (usually 8) of load or load group contacts between "on" and "off" according to programs stored in the microprocessor's memory. Battery packs may be used as protection against power failure. Functions often provided include:

Night Setback: Night override switches can be set to activate and deactivate connected heating and air handling units as required to maintain a minimum, preset night temperature.

Optimized Morning Start-Up: Early morning start switches can be set to optimize connected HVAC equipment start times. The user dials in an extreme summer lead time and an extreme winter lead time, and the programmer automatically computes the lead time for other mornings, based on actual outside temperature. The outside damper control remains closed until occupancy.

Adjustable Delayed Shutdown: Delayed shutdown switches can be set to permit selected (or all) connected loads to remain on for a user-prescribed additional length of time, after the scheduled shutdown time.

Holiday Switches: Holiday shutdown switches permit manual switching of the system into the holiday (night) mode, for any day of the week.

Optimized morning start-up, adjustable delayed shutdown, and holiday features generally can be activated or deactivated at any time, without altering the original programs stored in the system programmer's memory.

Some units can be reprogrammed by using one switch to deactivate all output signals, temporarily deactivating the night setback, optimized morning start-up and delayed shutdown features, and switching program/run switches to the "program" position. As the clock is advanced through the week, the program input switches are flipped back and forth between "on" and "off" states at the appropriate turn-on and turn-off times. Indicator lights confirm that the desired program is actually being stored in the memory. When programming is complete, the four program/run switches are returned to their "run" position; special features and outputs are reactivated.

OPTIMAL START/STOP CONTROLLER

As noted above, optimal start/stop is typically one function of a Level I controller. However there are also discrete controllers on the market which perform only this function.

An optimal start/stop programmer starts HVAC equipment when it is actually needed. The programmer works by comparing outdoor and indoor air temperatures. The difference between the two indicates when the HVAC system must be started or stopped in order to meet conditions desired when the first building occupants arrive, or when the most heavily staffed shift begins. The logic system of the controller is preset to take the building envelope's heat transmission factor into consideration.

Equipment is available representing various degrees of sophistication. For example, some systems rely on dry bulb temperature only, while others consider total heat content of both indoor and outdoor air. The optimal stop function is essentially the same as optimal start, except it stops HVAC equipment when prevailing indoor conditions are sufficient to meet the needs of the building for the remainder of the occupied period.

DEMAND CONTROLLER

Most building owners/managers receive an electricity bill that consists of two basic charges: one for the total amount of electricity (in kWh) consumed in the billing period; the other for electrical demand (in kW) representing the maximum amount of electricity consumed in any given demand interval during the billing period.

The demand bill is justified in terms of the electrical utility's investment needed to meet instantaneous demand. In other words, if a facility has a maximum demand of, say 1,500 kW, the utility must be capable of meeting that demand at any given time. Even if that demand actually occurs only once per billing period, the utility's standby generating capacity and equipment for transmission and distribution must be ready to meet it all the time. Thus, while two facilities may consume the same amount of electricity in a billing period, if Facility A consumes its electricity in 200 hours compared to 400 hours for Facility B, Facility A will have a much higher demand.

The length of the demand interval and associated rate structures vary considerably from utility to utility. In some cases it can be as long as 30 minutes; in others, as short as five minutes or even on an instantaneous basis. Demand charges may also be affected by a so-called ratchet clause. In that case, the billed demand may be no less than a certain percentage of actual demand experienced during a certain earlier period. In other words, for example, if actual demand in September is 200 kW, and a ratchet clause in effect states that billed demand cannot be less than 70% of the maximum demand recorded during the months of May, June, July, or August, the billed demand would be 210 kW, if during one of the summer months the peak demand recorded was 300 kW (300 kW × .70 = 210 kW).

When a building reduces its maximum demand, it helps the utility save energy and reduce its requirements for new generating capacity, since less energy is needed when the instantaneous demand the utility must be prepared to meet is reduced. The building's energy consumption is reduced, too, but only to a minor degree. However, the building's demand bill can be reduced to a significant degree, in some cases resulting in annual savings of thousands of dollars. In fact, if nothing has been done in a facility to limit demand, chances are that application of demand management will have immediate positive impact.

Note that demand management is not the same as demand control. Demand management involves two steps. The first involves a survey that identifies when peak demands are recorded, and the equipment operating at the time which causes that demand. Through study of these and related data, it may be possible to simply reschedule certain operations, so demand is lowered. For example, if three pieces of equipment are started simultaneously, the demand would be far higher than that recorded if equipment start-up was staggered, so that the second and third pieces of equipment were started at, say, five-minute intervals.

The second element of demand management is demand control, where a demand controller is used to take certain loads, called secondary loads, off-line, to prevent the present demand limit from being exceeded.

In typical application, a demand controller is connected to the utility's demand meter. Through use of any one of several different demand controller logic systems, discussed below, the control device determines electrical consumption during a demand interval and sheds the secondary loads, as needed, to ensure that the present demand limit is not exceeded.

Figure 5-1 is a graphic representation of demand during fixed demand intervals in a hypothetical building. With demand control in effect, the peak shown in the 8:15–8:30 interval would have been lowered. But the procedure of lowering the demand is creating problems for some utilities.

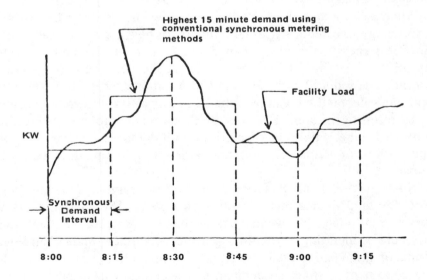

Fig. 5-1. Highest Demand Using Conventional Synchronous Metering

Because demand intervals are fixed throughout a utility system, widespread use of demand control equipment by utility customers creates an undesirable condition called "roller coaster effect." The utility experiences a systemwide high demand at the beginning of the interval, and a lowering of demand near the end of the interval.

In order to avoid this situation, some utilities are no longer permitting users to tie in with the utility meter, to obtain the end-of-interval pulse that is required. Some utilities are taking it a step further, by doing away with fixed intervals altogether. They are resorting to a "sliding window" approach, which is illustrated in Figure 5-2.

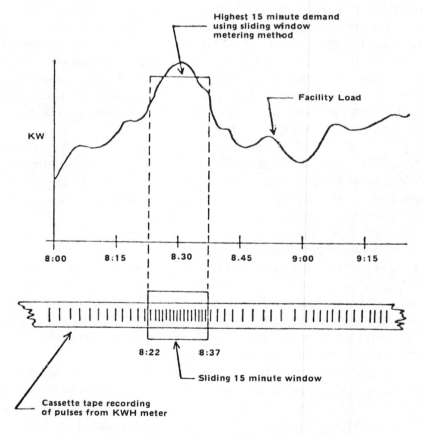

Fig. 5-2. Highest Demand Using Sliding Window Metering

This approach is applied through the use of advanced demand metering equipment which utilizes a magnetic tape to record consumption data. At the end of a billing interval, the tape is removed and a clean one is installed. The tape that is removed then is scanned with computerized equipment to find that consecutive 15 minute (or other) period during which maximum demand is recorded. In other words, the demand interval can start at any time. This approach also results in higher demand being recorded. This is illustrated by comparing the graphic representation shown in Figs. 5-1 and 5-2. In Fig. 5-1, the peak demand occurring 8:22 and 8:37 is split between two fixed demand intervals. By contrast, when the sliding window approach is used, the interval "starts" at 8:22.

In the event that a sliding window approach is used, or if the utility does not allow connection to its demand metering equipment, or if the utility indicates that it may in the future disallow connections to its demand meters, it is best to tie demand control equipment directly into the utility's incoming lines, using current transformers for the purpose. Note that some demand controllers are not able to function when a sliding window approach is used.

The trend toward use of the sliding window approach greatly increases the value of effective demand control, since the sliding window approach will increase demand charges. But the sliding window approach is just one of several approaches being used, some simultaneously, that also will increase the demand bill.

One of these new approaches is called "time of day metering." Through this approach, a day may be broken into two, three or even as many as five different periods, for example, 6AM–6PM, 6PM–12 midnight, and 12 midnight to 6AM. The demand rate, as well as the energy rate, will differ for each period. It will be highest during periods of maximum systemwide use; lowest during periods of minimum systemwide use. Through demand control and rescheduling of certain operations, demand charges can be minimized.

Still another aspect of demand and energy billing procedures relates to seasonal rates. Because systemwide demands generally are highest in the summer months, "summer rates" are higher than winter rates. Again, demand control can help in keeping charges to a minimum.

There are a number of demand controllers available. Their operations differ on the basis of the logic system employed.

The five logic systems generally used are: ideal rate, converging rate, instantaneous, continuous integral, and predictive.

Most systems use either one type of logic or another. Some systems use variants of each, or use two logic approaches and/or variants at the same time. Most systems will accommodate up to 64 loads. Some loads comprising numerous small motors, pumps, etc., can be connected in series and used as one load.

Ideal rate demand control logic accumulates and compares actual rate of energy use in any demand interval to a predetermined, ideal rate of use. As the actual rate of consumption approaches the ideal rate, loads are shed. As the consumption rate declines, loads are restored.

The ideal rate of energy use selected assumes that energy will be consumed at a constant rate. Because a true constant rate consumption pattern would begin at zero, the ideal rate begins with an offset. (If there were no offset, there would be no difference between the ideal and actual rate at the beginning of each demand interval, which would result in automatic shedding far too early in each cycle.) The amount of offset generally selected is approximately equal to desired maximum demand less energy consumed by primary loads at maximum usage. Shed and restore lines also are selected when establishing offset, to reduce the amount of on–off exercising which otherwise would be required.

Converging rate demand control logic is essentially similar to ideal rate logic except the ideal use curve and shed and restore lines converge at the end of each interval rather than run parallel to each other. This modification to the ideal use method provides a finer degree of control (usually over relatively small loads) late in the demand interval when accumulated registered demand may be very close to the maximum desired demand setpoint.

Instantaneous demand control logic compares actual rate of consumption at any time in the demand interval to maximum desired consumption at any time. For example, assume that maximum desired consumption is 1000 kW for a demand interval. With straight-line accumulation or constant usage, the accumulated demand should be no more than 250 kW at one-fourth of the interval and 500 kW at one-half of demand interval. Loads are shed and restored according to this criteria.

While this particular method generally is more accurate than the ideal rate method, depending on the width of the control band, average end-of-cycle consumption will usually be lower than maximum prescribed demand. Use of this approach also implies a certain degree of load equipment short-cycling. In cases of problem equipment, however, impact of short-cycling can be dampened through use of cycle timers on the equipment's control circuits.

Continuous integral demand is similar to instantaneous demand control in that usage is continually monitored and compared to the desired maximum rate. When actual rate approaches maximum desired rate, a load is shed, but the actual shedding function is carried out by a satellite cycle timer which deactivates the load for a fixed period of time. This results in demand interval overlap. For example, if a six-minute secondary load is shed at 14 minutes into a 15-minute demand interval, it will stay shed for the first five minutes of the following interval. Continuous integral demand controllers compare the present usage in previous intervals without restoring all loads at the end of each interval. This type of logic reduces excessive short cycling of equipment and operations.

Predictive demand control logic, generally the most popular now in use, periodically measures the total accumulated energy consumption from the beginning of a demand interval. A predicted or projected end-of-interval demand is developed based on rate of accumulated consumption measured at each set point and the rate multiplied by the time remaining in the demand interval. If the predicted demand exceeds maximum desired demand, a secondary load is shed. If a subsequent measurement indicates that maximum desired demand still will be exceeded, another load is shed. If a later measurement demand indicates a large enough margin, loads are restored.

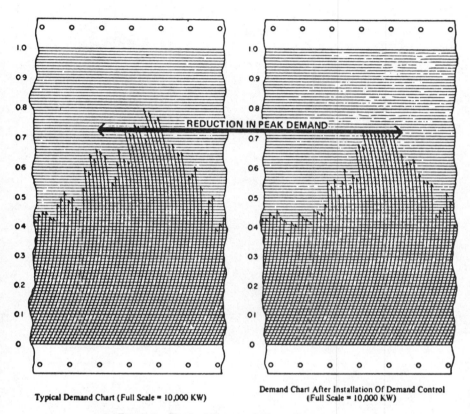

REDUCTION IN PEAK DEMAND

Typical Demand Chart (Full Scale = 10,000 KW)

Demand Chart After Installation Of Demand Control
(Full Scale = 10,000 KW)

**Fig. 5-3. Typical Demand Chart Compared
After Installation of a Demand Control**

6

Level II EMC System: General Monitoring and Control System

Level II EMC comprises a central monitoring and control system operated from a single command console. System functions are provided through a combination of electro-mechanical, pneumatic and extremely limited computer based control. For the most part, these systems have been outmoded, and only found in existing facilities. However, many of these were upgraded in recent years by energy-conscious engineers and building owners seeking to improve efficiency.

The Level II systems described in this chapter are less prevalent with state-of-the-art HVAC equipment, but some manufacturers still use similar systems. These manufacturers incorporate hybrid systems into their equipment with microprocessors to augment some control functions, and replace part of the relay logic used for conventional systems.

Monitoring is defined as the gathering and transmission of information, primarily that which relates to the status of the HVAC equipment and the spaces being served. Other systems can be connected as well. Monitoring generally is considered to include alarm, which is an audible, visual, or audible and visual notification that some element or a system or space is in an abnormal state. Typical system parameters include temperature, humidity, pressure, fluid flow, operating status of fans, pumps and similar devices, presence of fire and smoke, operating time, noise and vibration, presence of authorized or unauthorized personnel, etc.

Control is defined as a response to information obtained through monitoring, such as starting and stopping equipment, adjusting valves and dampers to control fluid flow, resetting set points of local controls, speed control, and transmitting visible or audible commands.

Control functions can be performed either manually or automatically. However, because Level II EMC systems employ hardwired logic for automatic control, when provided, the automatic functions they can perform are limited. Changing programs is extremely difficult; employing user-programming is impossible.

A trained operator is responsible for the efficient operation of a total system based on his own decision-making capability. Therefore, this type of system is totally dependent upon the operator for optimum system operation. This system can be adequate depending on the size and sophistication of the system being controlled.

A Level II system may be capable of status, analog and control point alarm indication, remote analog set point adjustment, and remote start/stop. Status and analog indication at the central console provide information on building conditions and can notify the operator of abnormal conditions in the system before they occur in occupied spaces. The remote start/stop function allows the operator to control remote equipment from the central console and thereby provide a prime manpower savings by reducing manpower requirements. It should be noted that for small, less sophisticated systems, the operator can perform optimization functions on the system by manual inspection and control of specific remote control points.

As a rule, Level II systems are not specified today as they make little economic sense. They cannot provide optimizing functions in a cost-effective manner, and are typically retrofitted in the field through the use of a Level I system, or a Level III system.

In the event that a building already has a Level II system installed, several alternatives for upgrading are available. The most drastic and costly of these alternatives is to remove the Level II system and replace it with a Level III system. This step should not be taken in any case, unless and until other alternatives are studied. One of these, as already mentioned, is the possibility of installing Level I control devices which could, if desired, be connected to the existing Level II system. Another alternative would be to upgrade the existing Level II system to Level III, by salvaging as much of the original equipment as possible, then adding or substituting upgraded equipment as appropriate.

Since many of today's existing Level II installations are packaged systems, provided by a sole manufacturer, and since upgrading may be performed by a provider other than the sole manufacturer, consideration must be given to the impact which upgrading would have on any remaining warranties and guarantees. Although the original provider would be in a position to extend the original warranties and guarantees, others can sometimes perform the conversion for less cost. Thus, while the original manufacturer should be asked to provide a cost proposal, several others should be asked to do the same. Costs and related benefits associated with having a new provider perform the upgrading may be such that it would be worthwhile to forego whatever remains of original warranties and guarantees.

7

Level III Automation System Technology

A Level III Automation or EMC system is also referred to in this book as an engineering control center. This is because these systems perform a central engineering function, similar to the concept of a Level II system, along with automatic control, energy monitoring, maintenance scheduling, information management, life safety and other functions which are made possible by a software-based logic system. The current state of the art in this technology is the Direct Digital Control (DDC) system. DDC is treated separately in Chapter 10. A further enhancement is distributed processing which will be discussed later in this chapter.

Level III systems can be divided into two specific "Hardware Types" based upon the equipment in use: Microprocessor- and Mini-computer-based systems. Level III systems can be divided into two "Function Categories" as well: Packaged and Hybrid.

HARDWARE TYPES

A *Minicomputer-Based System* is a high cost, sophisticated piece of equipment, typically requiring a dedicated control room and defined space temperature and humidity. These systems have often been associated with having the largest point capacity and system power. Until recently, this distinction also included the dividing line between systems providing basic control, and those with more sophisticated options such as: analog outputs, HVAC optimization, fire and life safety features, etc.

A *Microprocessor-Based System* is typically housed in a self-contained enclosure and operates in a wide range of ambient conditions. Developments in the computer industry have expanded the capabilities of these systems substantially. Today this type of system can provide many sophisticated control options, and at a low cost. Managers are finding that fewer automation dollars can purchase greater sophistication, flexibility and capability with these systems.

In preface to individual discussions of Level III system types in Chapter 8, this chapter focuses on technology issues. Advances in automated building control have developed through the availability of affordable computer technology. This chapter will discuss the basic concepts of this technology, to lay a foundation for examining Hardware and DDC issues in the following chapters.

FUNCTION CATEGORIES

A *packaged system* is one which is complete in every respect, furnished by a manufacturer who assumes sole responsibility.

A *hybrid system* is one which is composed of elements fabricated by several manufacturers. Depending on the nature of the contract, one vendor may assume sole responsibility for the entire system.

A packaged system, by its nature, is somewhat limited. If it does not satisfy all needs by design, very little can be done. A hybrid system, by contrast, can be designed to fit needs precisely, and very often can accommodate the very latest state-of-the-art components.

Hybrid systems are also more sensitive to advances in design concepts such as distributed processing, which will be discussed in this chapter. These systems allow users to acquire desired features at a lower cost. Discrete systems for energy management, fire and life safety may provide quality function, and yet have a lower first cost than a more sophisticated packaged system. Technology advances discussed in Chapter 11 on communications, and in Case Study 4, provide examples of hybrid systems.

The following discussion relates to the hardware and software associated with Level III EMC systems. A review of specifications for a Level III EMC system is provided in Chapter 13.

It is essential for managers working with the development and management of a Level III EMC system to be familiar with the concepts and "language" of electronic data processing. The following materials provide some of that information. Additional reading and other educational pursuits in the field are suggested.

COMPUTER

Computer is a general term used to define an assemblance of components which can be broken into two major elements: CPU (central processing unit) and main memory.

The computer serves to process information extremely rapidly. The first computers developed were able to perform in 20 minutes calculations or computations which otherwise would have required 100 mathematicians working for 100 years. Today's computers can perform in excess of one million such calculations in just one second. All of them will be performed accurately, providing that the persons operating the computer perform their functions properly; thus the well-known computer-era expression, "Garbage in; garbage out."

COMPUTER FUNDAMENTALS

The basic functions of a computer can be summed up as IPSO, for:

1. Input
2. Store
3. Process
4. Output

The four functions are performed on the basis of information called data. Computers work basically in the same manner as a pocket calculator, many of which also can perform the IPSO functions. The major difference is that, with most calculators, each step input, store, process, and output must be performed by hand. With a computer, each of the four functions is performed automatically in certain predetermined sequences, according to a specific program.

Today, however, even some calculators come with programs making them, in essence, hand-held computers. Accordingly, the major difference between a sophisticated computer and a sophisticated calculator is that the computer usually can receive more data, and from many more different types of input devices. A computer also can store more data, process the data in more ways, and display the data on a larger variety of output devices.

Fundamental to the speedy and efficient operation of most modern computers is the utilization of highly miniaturized integrated circuit (IC) components. These are manufactured by a highly sophisticated method known as large-scale integration (LSI).

Entire IC's are so minute that one can be placed on a single silicon wafer. Further, these wafers (or "chips") may contain hundreds of logic gates and flip-flops internally interconnected. Each chip performs a designated function within the computer. For instance, a chip can contain a number of registers or a complex decoder circuit.

Integrated circuits offer significant advantages over earlier computer components: miniaturization; reasonable cost; low power requirement; high-speed execution of logic operations; low heat dissipation. Most EMC systems rely on digital micro- and/or minicomputers. However, if a larger capacity is required, medium-to-large-scale computers also are available. Table 7-1 indicates distinguishing characteristics of various computer types.

Table 7-1. Characteristics of Various Computer Types

SIZE	LARGE Large Room	MINI 19" X 8'	MICRO 3" X 3"
Instruction Time	1 nsec	200 nsec--2 nsec	.5 nsec--15 nsec
Clock Rate	60 MHz	10 MHz	5 MHz
Instructions		70–200	24--158
Memory	262K	128K–1024K	1024K
Applications	Number Crunching	Dedicated Processing	Dedicated Control and Dedicated Processing
Word Length	Up to 60 Bits	8 to 32 Bits	1 to 16 Bits

Computers can also be categorized as synchronous or asynchronous. All the internal activities of synchronous computers are synchronized with an internal clock which emits several million pulses a second. Each activity of the central processing unit (CPU) is assigned a maximum execution time in pulses (machine cycles). During execution of an instruction, clock pulses trigger the next operation after the previous time interval. The CPU's control logic then waits for the next clock pulse.

Asynchronous computers do not delay CPU functions, because they are event driven. When an operation is completed, a signal initiates the next operation. Asynchronous computers do not require a master clock, but more hardware is needed to keep track of the status of CPU operations.

Units of Information

Each basic computer component has only two operating levels: on and off. As such each component can represent only two values: 0 (for "off") and 1 (for "on"). When a digital computer is operating, therefore, each one of its thousands of components can represent only either a 0 or a 1. Accordingly all computer information must be converted to a specific combination of 0s and 1s so the computer can recognize and handle the information.

Bits, Bytes, and Words

Digital computers perform calculations by counting digits. When only two digits are used, the term is binary digital (the prefix "bi" indicating "two.")

The smallest unit of information which a computer can work with is a *binary digit*, meaning one digit of the two possible—0 or 1. The term binary digit is abbreviated as a *bit* (from binary digit). Accordingly, a bit is always either a 0 or 1.

A computer cannot store or retrieve individual bits when transferring information to or from memory. Rather it works with larger elements of information called *"bytes"* and *"words."* Both "bytes" and "words" contain numerous bits.

Bits which are grouped together and handled as a single unit of information are called a *word*. Each word consists of a specific number of bits. Digital computers are word-oriented. They store individual words in memory and they retrieve individual words from memory.

The significance of a computer word depends on the computer being used, the size of the word, and the specific conventions employed by the user. A single word may represent one decimal number, or three decimal numbers, or three letters of the alphabet, or a few special symbols, or even an actual English word.

The number of bits in a word depends on the size of the digital computer involved. A minicomputer word may be as small as 8 bits, while a large-scale computer word may contain 60 bits.

Since computer information may be either instructions or data, computer words represent either instructions or data, and so are called *data words* or *instruction words*.

In some cases, words are divided into smaller units called *bytes*, which usually are either one-half or one-quarter the size of a word.

Like words, bytes contain a fixed number of bits and are treated as a single unit of information. Bytes are used when it is easier and acceptable to work with a unit smaller than a full word, especially if the word contains a large number of bits.

The number of bits in a byte depends on the size of the computer word. For example, a 12-bit word could be divided into two 6-bit bytes. Because most bytes are either 6 or 8 bits long, they are used to represent quarter-word lengths only when larger words are used. For example, a 32-bit word would normally be divided into four 8-bit bytes. An 8-bit word would not be subdivided into bytes at all.

Bytes, also like words, may consist either of data or instructions, and they may be stored in/or retrieved from memory. *The byte is the smallest unit of information that can be addressed.* Not all computer memories are byte-addressable.

In general, the longer the word length, the greater the efficiency and accuracy of a computer's internal operations.

Most of the minicomputers currently on the market have a 16-bit word length, divided into two 8-bit bytes (Figure 7-1). Other minicomputers have word lengths of 8, 12, 18, 24, or 32 bits.

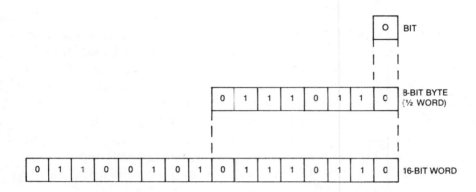

Fig. 7-1. Relationship of Bit, Byte, and Word

Units of Time

A typical digital computer performs one million or more additions per second. Accordingly, individual operations are measured in fractions of a second.

The most common units used are *milliseconds, microseconds,* and *nanoseconds,* as indicated in Table 7-2.

Operations which occur in input units, such as a teletypewriter, or output units, such as a line printer, are measured in milliseconds. The millisecond is too long an interval to accurately measure CPU or memory operations, however; they usually are measured in microseconds or nanoseconds. For example, a computer might perform one complete addition in just 2.6 microseconds, while a CPU can locate and retrieve a word from memory in just 800 nanoseconds.

Table 7-2. Units of Time

Units of Time	Fraction of a Second	Abbreviation
millisecond microsecond nanosecond	thousandth (1/1000) millionth (1/1,000,000) billionth (1/1,000,000,000)	ms or msec μs or microsec ns or nanosec

CENTRAL PROCESSING UNIT (CPU)

For a computer to solve a problem, it must have a sequence of instructions, a program which tells it precisely what operations to perform and where to find the data to be operated upon. The central processing unit, or CPU, is a specific unit which retrieves individual instructions from memory and then executes them as such. The CPU controls and supervises operations involving the other system units, and executes decision-making functions called for by the program. The CPU also performs all of the arithmetic computations.

The CPU is a microprocessor composed of two basic parts: *control unit (CU)* and the *arithmetic-logic unit (ALU)*.

The CU (Figure 7-2) coordinates and directs the activities of the entire computer system, by:

1. Locating and retrieving instructions from memory, one at a time

2. Decoding each instruction and generating control signals to start the specified action, and

3. Directing and controlling data movement in the CPU, memory, and input/output devices.

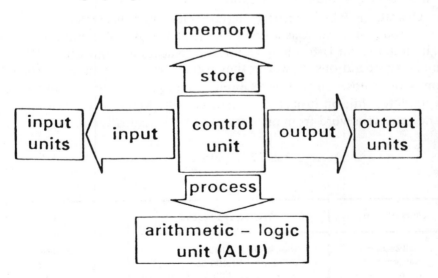

Fig. 7-2. Control Unit Functions

The ALU (arithmetic-logic unit) has two primary functions:

1. Performance of arithmetic computations, and

2. Performance of logical tests, such as the comparison of two values or a test for a zero value.

An arithmetic computation produces a result. A logical operation is used to make a decision. Note that all ALU operations are directed by signals from the CU (control unit).

CPU Components

The essential elements of a basic CPU are shown in Figure 7-3 and are discussed as follows:

1. The *Instruction Register* is a circuit that can electronically hold one word, so only one instruction can be executed by the microprocessor at one time. The sequence of instruction is usually stored in an external memory. As each instruction is needed, it is fetched from memory and put into the instruction register.

2. The *arithmetic/logic unit* (ALU) performs arithmetic and logic operations defined by the instruction set. It is the most critical element in the CPU.

3. The *accumulator* is the primary register used during many CPU operations. For example, it is used by the ALU to hold one element of data during arithmetic processing.

4. The *program counter* keeps track of where in the program the current operation is located.

5. The *index register* stores information that will be tapped by the program many times. It also holds addresses of reference data, such as tables of numbers.

6. The *extension register* is used with the accumulator to perform special arithmetic modes.

7. The *status register* is generally assigned to monitor the status of conditions within the microprocessor.

8. The *storage-address register* holds the memory address of data

that are either being loaded (read) from or stored (written) into memory.

9. The *storage buffer register* temporarily stores information. It acts as a buffer against surging of data flow into and out of the microprocessor.

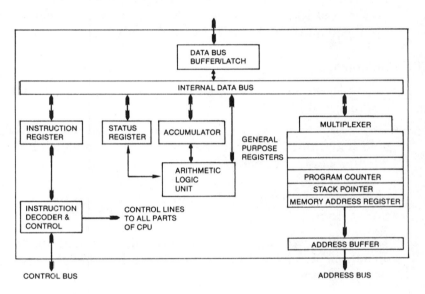

Fig. 7-3. Microprocessor (CPU) Functional Block Diagram

A CPU must be provided with the information that relates to the desired functions. It must be given instructions, and must be told where to obtain data to be operated upon, and where to put the result after the operation has been executed. The information transfer process is performed via communication channels that transmit a binary format into and out of the CPU. If the word length is eight bits, eight channels must be connected to the CPU. Collectively, these channels are called a bus. There are three types of busses.

1. The *data bus* transmits pertinent data into or out of the CPU especially to and from memory. The same bus usually is also

used for both reading data from memory into the CPU, and for writing data from CPU into memory.

2. The *address bus* is used to identify the individual location to or from which the transfer or data is to be made, especially in conjunction with memory. The bus carries the address of the location in memory that is being accessed by the CPU. It is also possible to address I/O (input/output) devices as if they were memory locations.

3. The *control bus* is a group of dedicated channels. Each channel can be used for a special control purpose, such as clearing CPU registers to reset the microprocessor, or stopping the device after instructions have terminated.

The interactions of these busses with the CPU, main memory and input/output is shown in Figure 7-4.

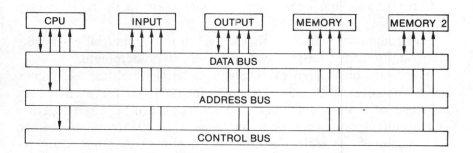

Fig. 7-4. Modularity and Bussing

MAIN MEMORY

A computer requires vast quantities of information. This information is stored in two areas: *main memory* and *auxiliary storage.* The data and instructions that are being processed by the central processing unit are held in numbered locations in main memory. Programs and data that are being stored for future use are kept in auxiliary storage. Programs and data can be transferred between main memory and auxiliary storage as often as necessary.

The capacity of main memory significantly affects the power and capability of the computer. The larger a computer's memory, the more instructions and information it can store and the faster it can perform.

Memories are available in many different sizes characterized by the total number of words that can be stored. Because only one location can be used to store one word, memory size also indicates the total number of locations in memory.

The letter "K" usually indicates the size of a particular memory. A "K" represents the value 2^{10}, which equals 1,024. Thus, a 1K memory contains 1,024 storage locations.

Although K represents 1,024, it is common practice to round off this value to 1,000. Thus, a computer memory with 4,096 locations would be referred to as a "4K memory." Memory sizes typically increase in multiples of 4K. Sizes 4K, 8K, 12K, 16K, etc., are common.

There are two basic types of main memory: *core* or *semiconductor.*

Core memory has been used for more than 20 years. It is fast, flexible, and reliable.

Semiconductor memories first appeared in commercially available minicomputers in 1970; today they appear to predominate.

A *core* is constructed of ferrite, a material which can be magnetized. Each core is about 1/20th of an inch in diameter (smaller than the head of a pin). More than 100,000 cores can be wired together and mounted on a single *array* or *plane.* Core planes are stacked to form a memory module. Core memory is nonvolatile; that is, any information stored in core memory remains unchanged when the power is off. Information can be changed only by passing a suitable current through the select wires. Also, core memory consumes very little power, and only during read/write operations (memory cycle).

Semiconductor memory has gained rapid popularity due to:

1. *Lower Cost:* Semiconductors store information electronically, enabling use of simplified circuitry which can be mass produced. Semiconductors come in simple, inexpensive, easy-to-assemble units that fit almost any new computer design.

2. *Higher Speed:* Semiconductors perform a read or a write cycle in just one operation.

The major disadvantage of semiconductors is their volatility. If there is a power loss, the information stored is completely lost.

Like other computer logic components a semiconductor memory unit is composed of numerous *integrated circuit* (IC) chips mounted on a printed circuit board. A typical semiconductor memory unit consists of semiconductor memory integrated circuit chips, location select logic, and read/write and control logic, all of which may be mounted on one or more printed circuit boards.

The *location select circuit* selects the memory chips which contain the bits of the referenced location. The *read/write circuit* performs a read or write operation. *Control logic* controls the entire sequence.

There are two basic types of semiconductor memories in current use: *bipolar* and *metal-oxide semiconductors* (MOS). The major differences between the two are: speed (memory cycle time), cost (price per bit), power consumption, and method used to hold information.

There are a variety of specific types of memory which use core, bipolar or MOS as their base.

Random Access Memory (RAM): can be MOS, bipolar or core. In this type of memory any word can be accessed without having to run through all stored words. Although RAM is volatile, it draws so little energy (typically less than ½ watt) that it can be left actuated at all times. Many manufacturers now supply a battery operated auxiliary power supply that is installed across the chips to counteract the effects of a power failure. RAM is a read/write memory, so the data stored can be easily modified or erased.

ROM signifies *read-only memory,* and can be either MOS or bipolar. It provides random access and is nonvolatile. Because the user cannot write into it, he must tell the manufacturer precisely what is wanted in the memory. ROM is often applied for storing tables, subroutines, and frequently used programs. As long as the information is never changed, it may be read as often as desired. Read only memories are also used in many computer systems to store the programs needed to start the computer when the power is first turned on.

PROM is a *programmable ROM.* PROM chips are purchased "blank" and are programmed to user specifications. Once programmed, the PROM behaves exactly like a ROM, and also is nonvolatile.

EPROM is an *erasable PROM,* giving the user a reprogramming capability. An EPROM can be programmed in the field, erased under UV light, or X-rays, and reprogrammed with different information. During reprogramming, the EPROM behaves just like a ROM. An EPROM is considerably more expensive but it is reusable.

EEPROM is an *electrically erasable PROM,* and is often called "E squared PROM." This chip also gives the programmer reprogramming capability; however it is not erased under UV light. Rather, EEPROMs may be reprogrammed via computer interface. These PROMs behave like ROM during reprogramming. However, with later versions it has become possible to simply modify the chip without complete erasure. EEPROMs are comparable in cost to EPROMs, but the potential exists for these chips to go down in cost, due to the elimination of the quartz window on the face of the chip.

CCD signifies *charge-coupled device,* which provides *serial access* rather than random access. Data is stored in the form of an electrical charge that can be moved from one memory cell to the next, like a pail of water moving through a bucket brigade. Because CCDs are accessed serially, they are slower than ROMs and RAMs, but they can store more data because they do not require address decoders. CCDs, like RAMs, are read/write devices.

Magnetic bubble memory also provides serial access. It is used for high-capacity read/write data storage.

PERIPHERAL DEVICES

Prior to describing these devices there is an issue of architecture to be clarified. A peripheral is a device used for storing data, entering it into or retrieving it from the EMCS. An architecture in this sense refers to the components which make up the system, and their proximity and interface with regard to one another. A variety of peripheral devices are utilized with Level III systems. However, the most prevalent of these today is not peripheral at all: the Personal Computer (PC). Though PCs may not appear to be germane to this topic, they must be treated due to their extensive use in lieu of the discrete peripherals discussed in this section.

The uses of the PC are particularly evident with microprocessor-based systems, but several users of minicomputer-based systems have begun to write their own PC interface software. To address these issues, this section will present the peripherals, including ancillary storage, as discrete pieces of equipment. It will then be followed by a brief overview of this function as fulfilled through a PC.

The use of discrete peripherals has traditionally been common with minicomputer-based systems. This is true for a number of reasons, but essentially there are three applications issues. First, these systems employ a central processor on site which has ample power to allow interface through peripherals, without the need for an ancillary computer. Second, these systems typically require specific ambient conditions, therefore a dedicated area must be constructed and conditioned. It is then convenient for that area to house both the EMCS and the control center. Third, these systems are often installed in large self-contained complexes, where remote communications is usually unnecessary.

It is important to note that regardless of the architecture employed, each of the major components is present in both PC and peripheral type systems. The distinction with the PC comes because an ancillary computer is driving the peripheral equipment, rather than the EMCS. A second distinction that is important to many users is that the PC may be used for other functions, such as office automation, etc., when it is not in use as an EMCS front end.

With systems that utilize discrete peripheral components, a number of options are available. As noted, peripheral devices are auxiliary input/output devices associated with the computer system. These include magnetic tape equipment, as well as auxiliary storage equipment such as magnetic disk devices. Many individual peripheral devices can function as combined input and output units and as auxiliary storage devices as well. Peripherals constitute a very important part of the computer system, and contribute substantially to the power of a computer, and its cost. The success or failure of a computer application often relates directly to the type of peripherals chosen. A thorough understanding of the advantages inherent in various types of peripheral devices is essential.

A discussion of some of the commonly used peripheral devices is provided below.

Keyboard: The keyboard is used to enter (input) information by depressing keys, as on a typewriter. The control circuit converts the depressed key into the equivalent binary code so the information can be utilized by the computer.

Printer: A printer prepares printed paper documents, and so is used for output only. Printers vary in speed, ranging from 10 characters per second to several hundred lines per minute. The format of printouts also varies. They may be on three-inch-wide paper tape or eight- or eleven-inch-wide (and wider) printed pages.

Two types of printers often are used. Typically, one printer records alarm conditions, and so is called the *alarm printer.* The other is a *data printer* which records logging information.

The alarm printer usually is arranged to record only those alarm conditions that users consider significant. Thus, at the beginning of each shift, an operator can identify the points that were in alarm, analyze the conditions involved, and take corrective action.

The data printer usually is keyed to record information that is normally logged by the operator during a routine shift. Typical logged conditions include chilled water leaving temperature, fan status and door security. Logging provides management with a summary of the status at the end of each shift.

Teleprinter: A teleprinter is a device which is combined with a keyboard and used as an output device. It looks and operates much like a typewriter, but internally it functions as two separate units: a keyboard and a printer. When the teleprinter is used on-line, the keyed information goes directly to the CPU. In turn, the CPU sends back (or echoes) the same character to the printers, if printing of the input is required.

Remote Terminal: An input/output device located away from the computer is called a remote terminal or station. Teleprinters are some of the most common computer terminals allowing computer users from various locations access to a single computer system. In this context, these stations are typically hard wired terminals.

Displays: A typical display terminal uses a cathode ray tube (CRT), a device which permits information from the CPU to be displayed on a visual panel. Some display terminals produce characters only, others can display graphs and pictures as well. Combined with keyboards, they are widely used in place of teleprinters.

Some older systems also provided a 35mm slide or filmstrip projector for display purposes. It consisted of an industrial-type slide projector and screen, along with slides of the various components of the system. In practice, when the operator realized, for example, that air-handling unit (AHU) No. 6 is moving into an alarm condition, a slide of AHU No. 6 is called for and projected manually or automatically, showing the configuration of the system, as well as the location and limit values of the alarm point in question. This enables the operator to analyze the problem and to take corrective action. Modern PC versions of this device include the concurrent display of point status or analog value by means of on-line communications. This means that an entire fan-coil unit or other system can be displayed, with values shown for return-air, mixed-air and supply-air temperatures; filter pressure drop; fan status, and so on.

The older style system displays also consisted of a combination of 35mm graphic displays surrounded by an array of several digital value displays and arrays of pilot lights to indicate operating status and alarm conditions.

AUXILIARY (BULK) STORAGE

Due to cost factors, main memory cannot accommodate all the data and instructions that the computer needs. Accordingly, additional information is stored in low-cost, large-capacity *auxiliary storage devices.* Auxiliary storage (also called mass storage or bulk storage) handles information in blocks containing hundreds of words rather than in single computer words that can be addressed individually. These large blocks can be moved rapidly into main memory whenever the computer requires them for processing. When the information is being processed in main memory, the CPU can address each word individually.

After the information is processed, the CPU can clear space in main memory by transferring information back to an auxiliary storage device. In this way, the computer need only accommodate, internally, a small portion of the information it requires to meet all of its data processing needs.

Auxiliary storage devices can be categorized by the way they access data, using sequential access or random access.

With sequential access, as mentioned above, data is stored in linear form, one after the other. If the data to be accessed is stored at the end of the storage medium, a relatively time-consuming process is involved.

With random access, any location can be accessed without reading through all previous locations.

Auxiliary storage devices are evaluated on the basis of three characteristics: data transfer rate, storage capacity, and cost. *Data transfer rate* is the time required to move information into or from main memory after the CPU makes the request. *Storage capacity* refers to the number of words that an auxiliary storage device can physically hold.

The basic media used with auxiliary storage devices are magnetic tape, magnetic disk, and floppy disk, discussed as follows.

Magnetic Tape: Magnetic tape is used frequently, for off-line storage. It is constructed of plastic and coated with iron oxide, which is magnetic. Magnetic tape has a number of narrow channels called *tracks* on which data is recorded. Large tapes may have as many as nine tracks.

Small electromagnets, called *read/write* heads, are located above each track, to read or write the data. Magnetic tape can be reused but the write operation is destructive. Accordingly, most devices have a "write protective ring." During the output process, the control circuit converts data from the computer into electrical signals. These are passed through the read/write heads and magnetize the surface of the tape. Once the write operation is complete, the drive mechanism rewinds the tape so it can be removed and stored. The process reverses during input. The control circuit receives data from the read/write heads and converts it to digital signals which are sent to the computer.

Magnetic tape devices can input and output data much faster than punched cards or paper tape devices. Large reels of magnetic tape can store millions of bytes of data and require less cabinet space than any other medium. Also, magnetic tapes can be changed and edited. In addition, they can be used interchangeably on the computers of different manufacturers, and so they are called compatible media.

Magnetic tapes are also used for auxiliary storage purposes in applications where high speed and random access are not required, and to store mailing lists and payrolls.

Cassettes: Tape cassettes are used to store smaller quantities of data. They do not have to be threaded through the tape drive and so are more convenient to use.

Magnetic Disks: The magnetic disk resembles a phonograph record. It consists of a magnetic surface that is rotated at a high, constant speed. Information is stored in circular channels called tracks which are further divided into segments called *blocks*. Each block can store hundreds of words of information. It performs both input and output functions, just as tape units, but it has the advantage of being random access. The disk drive mechanism spins the disk under an arm which contains the read/write heads.

Two types of read/write heads are available: fixed and moving. A fixed read/write head comprises numerous heads, one for each track on the disk. It is not removable because it would be difficult to realign them if they were disconnected to allow removal of the disk. This non-removable disk unit uses the same kind of circular magnetic surface as the removable disk unit. It differs only in the type of drive mechanism that transports information to and from the disk.

Fixed head disks are capable of faster reading and writing speeds than moving head disks, because the read/write heads are already aligned above each track, and the operation can be performed as soon as the disk rotates to the proper block. On moving head disks, the head must first be moved to the proper track and then the disk must rotate until the specified block reaches the read/write head. Where speed is an important consideration, fixed head disks are preferred. Because they are not removable, however, they cannot be used for input/output purposes and so are less versatile.

Like magnetic tape, magnetic disks can be altered and rewritten, and the write process is destructive. Unlike magnetic tape, magnetic disks are not necessarily interchangeable.

Floppy Disks: A floppy disk, or diskette, consists of a flexible Mylar disk, about 8 inches in diameter, which is permanently housed in a plastic envelope. Slots in the envelope allow the read/write heads access to the disk. It can serve as an input/output and/or random-access storage medium that is considerably smaller in capability and slower in performance than conventional disk units, and also far lower in cost. Floppy disks and floppy disk drive units are being produced by dozens of manufacturers and are finding their way into numerous small computer systems. Recent advances include floppy disks which provide more concentrated data storage, and double sided, double density floppy disks which utilize both sides, allowing more data to be stored on-line.

PERSONAL COMPUTERS

Rather than digressing into a thorough treatment of Personal Computers (PCs), this section will briefly discuss some key hardware and software issues. In particular these issues will be oriented to the EMCS Front End function. Chapters 11 on communications and 17 on program management deal in much greater detail with the software, and applications aspects, of the PC in relation to the EMCS.

Many users have found that the central station or control room concept does not suit their applications. As a result, remote communication is used extensively by managers for system interaction without leaving their desks. An EMCS Front End is defined as an ancillary computer system, typically a PC, with the capability of EMCS remote communications for monitoring, data access and programming of systems. This concept can be expanded to address such concepts as:

- Office automation functions—word processing, spreadsheet and data base management;
- Graphic presentation;
- Unattended or automatic status polling of multiple systems;
- Equipment maintenance scheduling and work order production;
- Management reporting.

For purposes of this discussion the basic definition will suffice. The emphasis in this section is not to explore futuristic features. Rather it is to acquaint users with the basic requirements for a PC front end. These will be covered in two areas: System Architecture and Hardware as well as System Compatibility.

System Architecture

In the section on peripherals, a number of components were examined. The term "architecture" refers to the combination of these components into a system, and is also used in discussing the individual components that make up an EMCS. The typical PC system used for front end interface consists of the following elements:

PC Cabinet
>Central Processing Unit
>Memory: RAM
>>ROM — and
>>Auxiliary Storage: Floppy Disk Drive(s) and
>>>Magnetic Hard Disk
>CRT Display — Color or Monochrome
>Keyboard
>Printer

A multitude of options is available to the users regarding processor speed, memory capacity, and display quality for both the CRT and the printer. It is essential for managers to carefully evaluate their short- and long-term needs prior to deciding on a PC. This process should be carried out with the aid of experienced computer professionals, and leads into the discussion of system compatibility.

System Compatibility

The criteria which is used by a manager and consultant in evaluating PC needs must address a number of critical issues. Perhaps the most important of these issues is compatibility. The standard of the industry has been set by International Business Machine's Personal Computer (IBM PC). Much of the software developed for use with the EMCS and for general applications is designed to operate with that system.

Hardware options—devices for communication and enhanced presentation—are also designed for the IBM PC. Therefore in assessing needs, managers must consider the software and hardware which will be required to accomplish current program tasks. Each item must be evaluated to ensure that the requirements for memory capacity, operating system compatibility, etc. will be met by the proposed system. In addition, the system should provide for future expansion to accommodate ongoing management needs. This is a complex function and should be approached with the same professional quality that is applied to making an EMCS decision.

FIELD EQUIPMENT

The field equipment associated with an EMC system comprises field interface devices (FIDs), multiplex panels, and sensors and actuators, discussed as follows.

Field interface devices may be intelligent or dumb panels. Intelligent FIDs are used with systems that employ distributed processing, and will be discussed under that topic in the next section. FIDs are central collection points where information from various sensors is gathered, sorted, coded and arranged for transmission to the central processing unit. They act as termination points for the remote ends of the transmission links and for connections to remote I/O devices in the area.

Dumb FID panels commonly include features such as: connection terminations for input/output (I/O) runouts and for transmission links; address storage and recognition logic; command interpretation logic; multiplexing logic; transmitter and receiver; power supply; transmission line interface; data-buffering and conversion capability; timing control, and modem. Many FIDs also include an analog-to-digital conversion unit, because the signal character of a sensor is normally of an analog or contact closure nature.

Each FID has one or more unique addresses which it must recognize and/or transmit when communicating with the CPU. It must also interpret commands from the CPU and perform appropriate requested transmission of input data and/or output functions.

MULTIPLEXER PANEL

Multiplexing is a data transmission technique whereby signals to and from numerous sensors and controllers attached to energized equipment can be sent over one line. The signals are coded to identify their source (when coming from a sensor or controller) or termination point (when being sent to a sensor or controller).

When the EMC system is being applied to one building only, multiplexing capability is included within FIDs. However, when multiple buildings are involved, sensors and controllers from several will be wired into a multiplexing panel, thereby reducing the number of FIDs required. This method reduces the first cost of the system; however, the quality of the devices used must be evaluated carefully to ensure control reliability.

SENSORS

Sensors are remote input devices which are connected to the terminal block usually located in the FID. Sensors measure the condition of a variable such as temperature, relative humidity, pressure, flow, level, electrical units and position of various mechanical devices. They also monitor relay, switch or other binary devices. The signal from a sensor or other input device is either analog or binary.

An *analog* signal is a continuously variable signal which bears a known relationship to the value of the measured variable. As an example, a thermocouple measures temperatures, and emits a signal in the form of a voltage; a resistance temperature device (RTD) measures temperature and emits a signal in the form of an electrical resistance.

A *binary* signal is an input signal equivalent to an electrical contact (switch) which can only be in an open or closed position, as determined by a predefined condition. Examples include firestats, door contacts, alarm devices, pressure switches, flow switches and motor-starter auxiliary contacts.

Typical analog and binary inputs to an EMC system are shown in Table 7-3.

In the past, for purposes of cost control, system manufacturers have specified "commercial quality" sensors. As a result, many of the sensors installed have been unable to provide the degree of accuracy necessary, particularly with analog sensing devices. Some are not UL listed.

Sensors must be selected with extreme care, despite the fact they

Table 7-3. Typical EMC System Inputs

ANALOG	BINARY (DISCRETE)

1. Temperature
2. Humidity
3. Pressure
4. Differential temperature
5. Differential pressure
6. Voltage
7. Watts
8. Amperes
9. KWH
10. Power factors
11. Electric loads/demand level
12. Steam consumption
13. Steam totalizing
14. Btu/totalizing
15. Btu/hr
16. GPM
17. Air flow
18. Enthalpy
19. PH levels
20. Contamination and impurity
      levels
21. Position devices (valves, dampers,
      etc.)
22. Process variables
	1. Alarms for:
	    a. Safety circuit alarm
	    b. Firestat alarm
	    c. High temperature alarm
	    d. Low temperature alarm
	    e. High pressure alarm
	    f. Low pressure alarm
	    g. Steam pressure alarm
	    h. Condensate level alarm
	    i. Liquid level alarm
	    j. Transformer alarm
	    k. Bearing temperature
	    l. No flow, flow failure alarm
	    m. Elevator/escalator malfunc-
	      tion alarm
	    n. Dirty filter alarm
	    o. Freezestat alarm
	2. Status of:
	    a. Motor operation:
	      i. On/Off
	      ii. Run/Stop
	      iii. Fast/Slow
	    b. Occupied/Unoccupied
	    c. Day/Night operation
	    d. Summer/Winter operation
	    e. Heat/Cool
	    f. Manual/Automatic
	    g. Open/Close
	3. Fire Alarm:
	    a. Supervisory flow switch
	    b. Smoke detectors
	    c. Heat detectors
	    d. Pull boxes
	    e. Zone
	4. Security Alarms:
	    a. Patrol tour call-in
	    b. Access detectors
	    c. Electric eye
	    d. Infra-red detectors
	    e. Motion detectors

are small and seemingly simple. In a fine-tuned system, a chiller temperature set point deviation of 2F will quickly be felt in all conditioned spaces. Likewise, unless binary signals used to sense fluctuations in temperature or pressure are absolutely reliable, logic and alarm inputs will be in error, resulting in excessive output errors.

Due to the fact that accuracy of sensing is critical to the performance of an EMC system, a brief discussion of commonly used sensors is provided.

Temperature Sensors

The most popular temperature sensors used for precise electronic control systems are resistance thermometers/RTDs, thermocouples, and thermistors.

1. *Resistance temperature detectors* have a temperature sensitive element, whose electrical resistance increases with increasing temperature in a repeatable and predictable manner. The sensing element is usually made of small diameter platinum, nickel or copper wire wound on a special bobbin, or otherwise supported in a virtually strain-free configuration.

   Resistance thermometers are usually selected for their high accuracy/stability and good sensitivity. A small electric current must pass through the resistance thermometer in order to produce an output signal proportional to temperature. Basic system accuracy depends on stable power source and proper selection of precision bridge resistors, which are invariant with temperature change.

2. *Thermocouples* basically consist of a pair of wires made from dissimilar metals. When the junctions of these two dissimilar metals forming a closed circuit are exposed to different temperatures, a net electromotive force is generated, which induces a continuous electrical current. The thermocouple measures only the temperature difference between its reference junction and the measuring junction. Accuracy depends on reference junction temperature compensation; typically electrical resistors, plus a temperature sensitive resistor, designed for the particular thermocouple type.

Thermocouples are usually selected for their high temperature suitability, for their inherent ruggedness, for their low cost in simpler forms. However, the cost of a thermocouple system is typically no lower than for a resistance thermometer of comparable accuracy, because the thermocouple has lower sensitivity, requires reference junction temperature compensation, and special extension leadwire.

For EMC system applications, thermocouples usually are restricted to a temperature range of 250F or above.

3. *Thermistors* are basically ceramic temperature sensitive resistors. Exhibiting high negative coefficients of resistance, these semiconductors possess resistance values which may vary by a ratio of 10,000,000:1, from −100 to +450C. (Thermistors for special applications can be used up to +500C and above and down to −180C or lower). They are available in a resistance range from ohms to megohms; their thermo-sensitive characteristics, coupled with stability and high sensitivity, make them a highly versatile tool for temperature measurement.

Thermistors are made of semiconducting materials, including a number of metal oxides and their mixtures, such as oxides of cobalt, copper, iron, magnesium, manganese, nickel, tin, titanium, uranium, and zinc. The oxides, usually compressed into the desired shape from powders, are heat treated to re-crystallize them, resulting in a dense ceramic body. Electric contact is made by various means—wires imbedded before firing the material, plating, or metal-ceramic coatings baked on.

The application of thermistors to the measurement of temperature follows the usual principles of resistance thermometry. Conventional bridge, or other resistance measuring circuits, are commonly employed. The high temperature coefficient of thermistors results in their having greater available sensitivity as temperature sensing elements than metal resistance thermometers, or common thermocouples.

To illustrate, the use of thermistors permits the adaptation of conventional temperature recorders to the measurement of 1C spans, which is not feasible with ordinary resistance thermometer, or thermocouple elements.

Because thermistors and resistance temperature detectors (RTDs) provide relatively high absolute accuracies, they are commonly specified for EMC application. A precision wound 100-ohm 4-wire platinum RTD is commonly used for temperature sensing. These sensors are typically specified to operate over the following ranges with the specified accuracies:

+ 20F to +250F	− accuracy ± 0.75F
+100F to +500F	− accuracy ± 2.0F
+250F to >	− accuracy ± 2.0F

Gauge Pressure Sensors

There is a variety of differential pressure sensors available on the market that provide electrical output. These sensors are known as *transducers,* differentiated by the type of sensing elements—bourdon tube, bellows, solid-state Piezoresistive, or diaphragm used.

Differential pressure changes the shape of the pressure element, causing a change in the electrical characteristics of the transducer, proportional to the differential pressure.

Depending on the medium being controlled, the sensors are usually specified to have the following pressure and accuracy ratings:

1. Chilled, condenser, and hot water sensors are rated at (125) psig.

2. Low- and medium-steam pressure sensors shall be rated at (150) psig. Low pressure sensors operate from 0 to 30 psig, with an accuracy of plus or minus 2%. Medium pressure sensors operate from 0 to 100 psig, with an accuracy of plus or minus 2%.

3. High steam pressure sensors are rated at (300) psig, have a full operating range of 0–300 psig, with an accuracy of ±2%.

4. High temperature water sensors are rated at 600 psig and have a full operating range of 0–400 psig, with an accuracy of ±2%.

5. Sensors on all steam lines and high temperature water lines are protected by pigtail siphons, installed between the sensor and

the fluid line, and have an isolation valve installed between the sensor and pressure source.

Relative Humidity Sensors

The sensors used to measure relative humidity are of resistance type, i.e., employing the principle that electrical resistance of materials changes with changes in the amount of moisture contained in these materials.

The sensing element specified is rated for the relative humidity range designed into the building environmental control system. The sensor should have an overall accuracy of ±2.5% of span over the range of 20 to 80% relative humidity.

Vibration Sensors

Vibration sensors specified with EMC systems usually have a range of 2 to 500 hertz and a field adjustable velocity set-point of 0.3 to 3 inches per second. Sensors are arranged for devices upon which they are mounted.

Flow Sensors

Differential pressure transducers and Venturi flow tubes for flow sensing are utilized. The differential pressure transducer provides a 4-20 ma or 0-100 mv analog signal as flow varies 0-100% of range. The signal is completely linear for the measured range of the differential pressure. A method is to be provided to convert the pressure reading (a square root of actual flow) to a linear reading of actual flow. The accuracy of signal for EMC applications is usually specified at ±2% over a range of 10% to 100% of flow.

All assemblies have adjustable range and differential pressure settings and operate automatically when conditions return to normal.

Pressure sensors used with EMC systems usually are specified to have the following features:

1. Adjustable high and low limits.

2. Suitability for operation in an ambient temperature range of 30 to 140F.

3. Accuracy within 1% of full scale.

4. Static pressure taps for ease of inserting a calibration meter.

5. Maintenance, shock, vibration, and pressure exposed to surge of 25 psi above scale.

Dew Point Sensor

The dew point sensor (dew cell) often has analog type precision-wound RTD as the sensing element. The dew cell is usually installed in a duct and is powered by a separate source. Actual overall cavity temperature is sensed and represents dew point to an accuracy specified at ±2F.

Dirty Filter Alarm Sensors

A dirty filter alarm sensor basically is a differential pressure switch with field adjustable set-points. Contacts are rated for dry circuit application. The switch is a dual purpose voltage/ampere contact rated with single-pole, double-throw snap action.

Alarm Level Sensor

An alarm level sensor is nothing more than a contact sensor, located either on the motor starter of the equipment or on the level sensor itself.

Current Sensor

Current sensing devices are the current transformers with windings of appropriate .current values when reduced to 5 amperes; a 0-5 ampere signal is converted to a dc low level signal (0-100 mv or 4-20 ma typical) for analog transmission to the FID for conversion on a per point basis. Accuracy of these sensors is usually specified to ±2% of the monitored range.

Voltage Sensors

The voltage sensing devices usually are voltage converter type with the input voltage being fed into the converter. The overall accuracy specified is ±2% between 200 and 500 volts. The voltage converter is transduced to a 4 to 20 ma or 0 to 100 mv signal for analog transmission to the FID.

Digital Inputs

All digital inputs are from contact type sensors. A typical contact rating is 1.0 milliamps.

All sensors used must provide input in a manner which meets performance requirements of the central processing unit. CPU manufacturers should be consulted for exact requirements of sensors and methods of wiring.

ACTUATORS

Actuators are devices which perform control action at the remote point in response to central system command instructions. Typical actuators connected to FIDs include controlled motors, valve positioners, damper operators, switches, and relays.

As with input devices, the output signals to the actuators can be classified as either analog or binary. Typical outputs from an EMC system are shown in Table 7-4.

DISTRIBUTED PROCESSING

Distributed processing EMC systems employ smart FID panels which perform functions that are handled solely by the central processing unit (CPU) in conventional systems (Figure 7-5). These smart panels incorporate microprocessor and memory, such as PROM, thus serving as remote CPUs. They are capable of handling many or all of the functions that the CPU can perform (scanning, control for start/stop, optimization, etc., monitoring and in some cases self-diagnostics) associated with the points connected to them.

Table 7-4. Typical EMC System Outputs

ANALOG	BINARY (DISCRETE)

1. Remote setpoint (control point) adjustment of:
   a. Temperature
   b. Humidity
   c. Pressure
 	
2. Position reset of:
   a. Switches
   b. Dampers
   c. Pressure
 	
3. CCTV Camera Control
	1. HVAC:
	   a. On/Off
	   b. Run/Stop
	   c. Open/Close
	   d. Fast/Slow
	   e. Flow/No flow
	   f. Manual/Automatic
	   g. Heat/Cool
	   h. Occupied/Unoccupied
	   i. Day/Night
	   j. Summer/Winter
	   k. Intercom
	
	2. Fire
	   a. Alarm bells
	   b. Door holders
	   c. PA system
	
	3. Security
	   a. Door locks
	   b. Bells
	   c. PA system
	   d. CCTV Camera Control
	
	4. Intercom
	   a. On/Off
	
	5. Lighting
	   a. Control and programming (On/Off/Auto)
	
	6. Elevator/Escalators
	   a. On/Off
	   b. Run/Stop

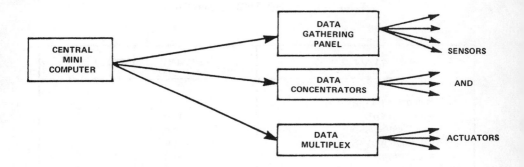

Fig. 7-5. Traditional EMC

First generation distributed processing systems were only available with minicomputer-based systems (Figure 7-6). In these systems functions which could not be handled by the so-called "smart" FIDs were then passed on through the system's hierarchy. Instead of going to the CPU, however, they were fed to an intermediary device called a Central Communications Controller (CCC), in essence, a stripped-down minicomputer. Only those functions which could not be handled by the CCC then were passed on to the CPU.

Today's state-of-the-art distributed processing systems are panels which serve as stand-alone controllers (Figure 7-7). The control algorithms for connected loads reside in these panels, often in ROM so that battery back is not necessary. In addition many panels are capable of self-diagnostic routines, sharing data and in some cases trending data. Network configurations of these smart FIDs allow communications and control via interaction with a master CPU, or other smart FIDs on a "Peer Bus Network." In this way a controlled load in panel A can be sequenced based upon input data from a sensor or device tied to panel B.

The primary benefit derived from distributive processing is greatly increased reliability. Because there is a great deal of system redundancy, one element of the system can take over for another when the need arises. This is particularly beneficial during performance of maintenance. Distributive processing also enables easier expansion, be it to integrate more points or to add more functions.

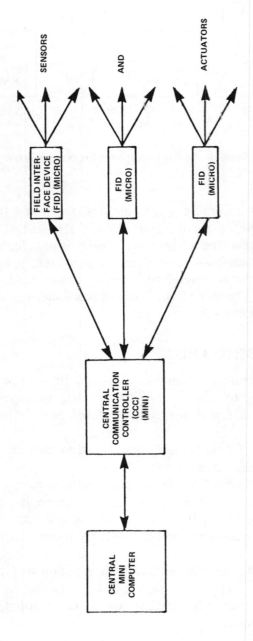

Fig. 7-6. First Generation Distributed Processing System

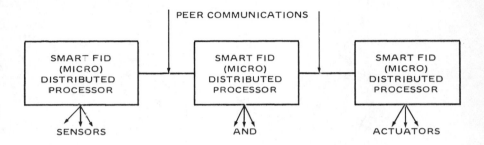

Fig. 7-7. State-of-the-Art Distributed Processing System

This concept is becoming more complex as smart FIDs are now being developed for Zone control of HVAC. The cost of these intelligent modules is reaching a level which makes it possible to conceive of a host of smart controllers operating independently, but interacting to share data and achieve global control. The discussion of Direct Digital Control in Chapter 10 is related to this issue, and will provide interesting expansion on the concept.

DATA TRANSMISSION LINKS

Data collected from and commands issued to remote points are sent through FIDs to and from CPU via data transmission links. They take many different forms depending upon project requirements.

Different types of data transmission links and their characteristics (Table 7-5) are discussed as follows.

Coaxial Cable: Coaxial cable is the most commonly used data link. Its multiplexing capability means that there is no practical limit on the number of facilities which can be connected to the system, making it an excellent choice, especially when expandability is a concern.

Coaxial cable can comply with any transmission or error rate requirements, because its transmission speed is very fast and error rate is very low. Such versatility makes coaxial cable compatible with any future requirements.

Table 7-5. Comparison of Characteristics for Several Data Transmission Links

	First cost	Scan time	Reliability	Maintenance effort	Expandibility	Compatibility with future requirements
Coaxial Cable	high	very fast	excellent	minimum	unlimited	unlimited
Triaxial Cable	high	very fast	excellent	minimum	unlimited	unlimited
Twisted pairs	high	very fast	very good	minimum	unlimited	limited
VHF FM radio signals	medium	limited application	low	high	very limited	very limited
Microwave	very high	very fast limited	excellent	high	unlimited	unlimited
Telephone pairs	very low	very slow	low to high	zero, if phone company does it	limited	limited
Fiber Optics	very high	very fast	excellent	minimum	unlimited	unlimited

Reliability is excellent provided minimum maintenance is provided.

Triaxial Cable: A triaxial cable has the same characteristics as a coaxial cable. It is composed of a coaxial cable plus an aluminum-mylar outershield and drain wire. It is used in cases where the cable will be exposed, that is, not run in conduit.

Twisted Pairs: A twisted pair is a dedicated, hardwired pair which is similar to telephone lines. Although its transmitting performance is similar to that of a coaxial cable, as is its expandability and maintenance requirements, nonetheless the overall capabilities of twisted pairs are not rated as highly as those of coaxial cable, often because data transmission rates are limited, depending on the particular kind of twisted pairs used.

VHF or FM Radio Signals: Although VHF or FM radio signals which start and stop functions are becoming popular, problems have been experienced.

The basic problem is obtaining the frequencies, due to the limited number available.

If frequencies can be obtained, interference can be a problem. An airplane flying overhead can trigger a system, a situation which can be avoided only through extensive fine-tuning.

Expandability also can be a problem. If 100 items are to be activated and deactivated at different times, 100 different frequencies and signals are needed.

Other problems include:

1. Signal distances are limited, since longer distances increase the possibility of interferences from airplanes, CB radios, etc.

2. High maintenance requirements due to the large number of transmitters and receivers involved.

3. Low reliability.

In some cases a combined system will be employed, whereby an FM radio signal and a carrier signal are carried on a power line. The carrier signal is transmitted through the low-voltage side of a transformer; the radio signal takes care of the rest of the transmission. A combined system is limited because of the large number of items normally transmitted on the low-voltage side. Scan time can be relatively fast, but there is a limit to what can be scanned.

Microwave Transmission: Microwave transmission is a practical alternative for communication between facilities separated by considerable distances. The positive attributes of microwave transmission include a fast scan rate which is compatible with most requirements; excellent reliability (assuming knowledgeable maintenance personnel are available), and compatibility with future requirements and expansion.

The primary negative attribute of microwave transmission is high first cost. Receivers and transmitters are needed in each building.

If microwave transmission is used, other types of data links should be specified for completing the trunk wiring system.

Telephone Pairs: Telephone lines are the most commonly used data link when the EMC computer is located remote from the building(s) served. See Chapter 11 for a detailed discussion of communications functions. The local telephone company charges a small initial connection fee and ongoing charges for monthly equipment lease. Maintenance is included in the monthly lease fee, with a certain level of service guaranteed.

There are many limitations to telephone lines.

Scan time is slow; rated in minutes rather than milliseconds. Scan time also is limited, usually to 1,200 baud, though scanning at 9600 baud has been achieved in some instances. By contrast, the scan capability of the central processor may be 50,000 baud or higher.

Reliability of the system depends on the type of equipment provided by the telephone company. Sometimes the equipment is antiquated.

The minimum quality of each line used in trunk wiring system must be clearly defined. The slow transmission rate of telephone lines may affect compatibility with future requirements. Likewise the high error rate, although initially acceptable, may not be so in the future. For example, if an extensive security system is added to the EMC later on, the error rate may be too high and the scan rate may be too slow.

If the telephone circuits are owned by a utility, they all must comply with tariffs of the Federal Communication Commission. These tariffs contain a complete and exact description of the services available, and are defined in terms of transmission rates, error rates, and other circuit characteristics such as amplification and damping.

When leased telephone lines are being considered, a question arises as to the use of two-wire or four-wire circuits.

With two-wire circuits, noise and signal distortion are fairly commonplace. While four-wire circuits eliminate these problems, they usually cost twice as much as two-wire and take twice the number of lines, which may not always be available. A multi-point circuit is sometimes specified to alleviate some of these problems. As shown in Figure 7-8 a signal flows to the CPU, travels to the central exchanger or out to a remote location, and from there to several different facilities. A bridge, or impedance matching device, is used to insure signal uniformity on all lines. Bridging is used on both two- and four-wire circuits, but it performs more easily on four-wire circuits.

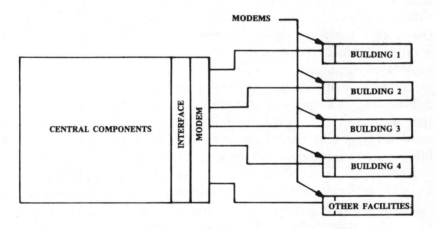

Fig. 7-8. Multipoint Circuit for Multifacility Arrangement

Figure 7-9 illustrates an alternative to the multipoint circuit, namely, a loop circuit in which telephone lines loop from building to building. Through this approach, some of the potential bridging requirements are eliminated, but telephone lines that loop from building to building generally are unavailable.

Two-wire circuits have been employed more regularly due to the expense of four wires and their lack of availability. In addition, Class A security or fire alarm systems require two paths of transmission,

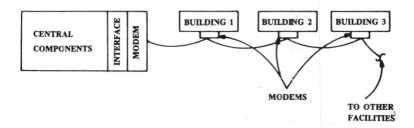

Fig. 7-9. Telephone Line Loop Circuit

meaning that eight wires are needed if a four-wire circuit is used. If an intercom is required, an additional pair of wires would be needed, bringing the total to 10, or five telephone pairs for each point of communication.

Carrier Signals on Power Lines: Although the use of carrier signals transmitted on electrical power lines has been accompanied by many complications, it could represent the way of the future. Since an electrical power distribution system already interconnects all buildings in a complex, initial costs would be extremely low. As noted, however, this form of data link is not without problems. For example, carrier signals on power lines cannot be transmitted through transformers, so a transmitter must be present to transmit and receive. Nonetheless, the theory and practicality of this method are feasible.

Hybrid Transmission System: Several types of data links can be combined into a trunk wiring system to develop a hybrid system. For example, a series of small motors within the system could be turned on or off via an FM radio signal. In another area, a coaxial cable can be used to handle certain high-speed functions. Telephone pairs could be used to communicate with remote buildings, more expensive coaxial cable for those which are closer. In such arrangements, each aspect of the system must be specified precisely to insure proper control and maintenance. Figure 7-10 is a schematic showing use of several types of data links in the trunk wiring system. If voice communication at remote panels is needed, an additional pair of telephone wires is required. Note, however, that hard-wired intercommunication is not mandatory; walkie-talkies may be adequate to meet needs.

Fiber Optics: Fiber optics represent a new technology which even-

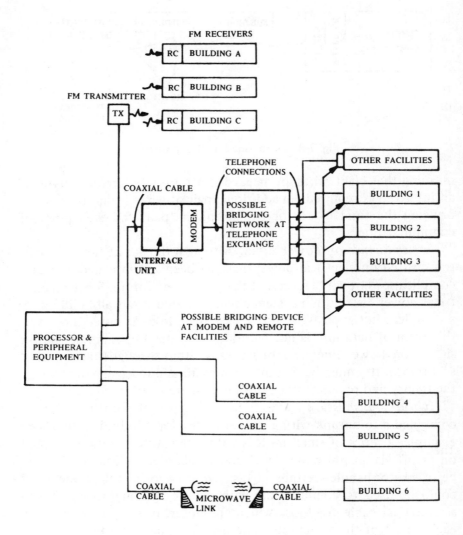

Fig. 7-10. Hybrid Transmission System

tually may replace the conventional wire and cable used in analog and digital communications systems.

With fiber optics, a smooth, hair-thin strand of transparent material conducts light with high efficiency. Because each strand must transport information independent of neighboring strands, a coating or cladding is applied to the fiber walls. The coating also protects the fiber from dirt and scratches.

The major advantage of fiber optics is its light weight and small size. Only 1½ lb of fiber optics can replace thirty pounds of copper wire. Fiber optics also offer freedom from interference, lower signal loss, and potentially lower system costs, depending on future developments. A single-fiber cable now costs about $0.30 per foot. Mass production is expected to lower cost to less than three cents per meter.

A fiber optic cable which contains 24 low-loss silica fibers in two ribbons of twelve fibers each, has an information capacity which exceeds 44.7 megabits per second. (In other words, if a light source were switched on and off 44.7 million times per second, the cable would keep each flash separate.) In addition, a single fiber optic can carry up to 50,000 voice channels, as compared to 5,400 channels for standard coaxial cable.

A topic that is related to this discussion is data transmission protocols. This involves the structure of the information that is transmitted, and the types of equipment that are required to receive the data. There is a movement in the EMCS industry towards standardizing these processes, with numerous ideas on the best method to accomplish the task. The schools of thought on this issue are discussed in Chapter 11 on Communications and Protocol Standardization.

SOFTWARE

Computer programs are called software. Because programming in machine code is both time consuming and error prone, nearly all computer software is generated by people who write, or program, in various computer languages.

COMPUTER LANGUAGES

A computer language is a set of specific words and phrases that allows people to tell a computer what to do. Each language has special rules and punctuation symbols for combining these words into instructions or statements.

There are many different computer languages. Each is designed to facilitate the process of transforming an algorithm into a computer program. Note that the *choice of the language by which programmers implement an algorithm frequently determines the computer's efficiency in a given application.*

Computer language should be selected on the basis of certain factors which divide the languages into two broad groups: low-level (assembly) languages and high-level languages such as COBOL, FORTRAN, and BASIC.

In the past, many users did not specify software programming languages. As a result, vendors selected the languages. Using a language chosen by and unique to a vendor can present difficulties. If it is decided to change vendors, the entire software investment may be lost, because the programs cannot operate on any other system.

Low-Level Languages: The most fundamental low-level language, and the only one directly usable by a computer, is binary machine code. Programming directly in machine code can be awkward, because the programmer must remember complex binary patterns representing each operation (operation codes) and the binary address of memory locations used. A more viable low-level language is an assembly language which uses mnemonic codes to reference memory locations.

There are four important characteristics of assembly languages:

1. Mnemonic op codes substitute for the binary op codes of machine code in assembly language. This eliminates the need to remember bit patterns or octal numbers for operations, because it substitutes easily remembered words or abbreviations, such as ADD for addition or STR for store.

2. The binary address of machine code are replaced by symbolic addresses in assembly language. Thus, the programmer must label only those addresses to which reference is sought.

3. Due to the mnemonic op codes and symbolic addresses, assembly language programs must be translated (or assembled) into machine code before they can be executed.

4. Each assembly language instruction corresponds to a single machine code instruction. Therefore, the assembly process is one of conversion to a single machine code instruction from each assembly language instruction.

A typical assembly language instruction contains up to four fields: label, operation, operand, and comment fields. Not all four must be used. The following is a sample assembly language instruction with four fields.

Label	*Operation*	*Operand*	*Comment*
START	ADD	NUMBER	/GET 1ST VALUE

Assembly language has these four advantages:

1. Assembly language is easier to learn.

2. Assembly language is faster to write.

3. Assembly language is less error prone. Even experienced programmers may produce errors when programming in machine code, for two reasons. First, the specification of the op codes and addresses must be precise. Second, checking and editing machine code may not be effective, due to similarities between instructions. In assembly language, the differences are magnified. For example, STR A might be 011 010 000 111 in machine code and ADD A might be 001 010 000 111. While STR and ADD are different in assembly language, in binary machine code there is only a difference in one bit between the instructions in the previous sentence.

4. An assembly language program is more easily documented, because it provides two important documentation aids not available in binary machine code. The use of labeled addresses allows data, as well as program steps, to have meaningful symbolic names. This feature, combined with the valuable comment fields, permits an assembly language program to be

documented externally and is nearly undecipherable without documentation.

High-Level Languages: High-level languages generally are tailored to favor the people who use them, rather than the computers on which the programs will be run. They are distinctly different from low-level languages.

High-level language statements (or lines) have a many-to-one correspondence with both machine code and assembly language instructions. Many of the elementary computer operations and hardware considerations are removed from the programmer's concern and control. Also, high-level language statements generally correspond closely to natural language or algebraic expressions encountered everyday.

High-level programs require the use of a translator, that is, a program which converts a source program into some other form such as object code or machine code. Assemblers are translators for low-level languages. There are two major classes of translators for high-level languages: *compilers* and *interpreters.*

Compilers: A high-level language source program is translated into object code by a compiler. This object code can then be linked into machine code. The complete translation process involves five phases.

First, analysis of the individual statements occurs. Every statement is checked for errors in obeying the rules of the language (syntax checking). A table that contains the simple machine operations corresponding to each statement is then generated. This table is frequently called "1-op table" because each line of the table contains only one operation.

The next phase of compilation is the generation of assembly language instructions from the lines of the "1-op table."

The third phase relates to the optimization of the generated assembly language instructions. The generated code was produced from the "1-op table." As a result, unnecessary store-fetch operations may have been created. These "wasted" instructions are removed during the optimization phase. Depending on the level of the compiler's sophistication, additional, more subtle optimizations may be performed. The optimization process is important; it enables high-level language programs to have execution speeds comparable to those of assembly language programs.

The fourth phase of compilation is the relocatable assembly of the optimized assembly language program. The object code produced by this phase is indistinguishable from the object code generated by a relocatable assembler or other compilers.

The last phase in the translation of a high-level language to machine code is the linking of the object code. Due to the fact that this phase is performed by the link, the procedure is not strictly part of the compilation process. The object code produced by a compiler is compatible with the object code produced by other compilers and a relocatable assembler. This allows the programmer to write individually each program and sub-routine in the best language for the specific application. The programmer then links the various object programs into a single machine code program that can be executed. Therefore, the programming development phase can be considerably shortened by careful selection of the languages.

Note certain important points regarding compilers:

1. The object code is the output of a compiler;
2. Compilation and optimization may be time consuming;
3. Because of code optimization, execution speed is relatively high;
4. Mixing suitable languages on a subroutine-by-subroutine basis, can reduce development time;
5. Development time is increased because logical errors in the program are not detected until completion of the compilation linking process. Therefore, it takes longer to find and correct a logical mistake;
6. A compiler's object code output can be saved for later linking and execution. Therefore, libraries of already compiled programs are easily created.

Interpreters: An interpreter produces the program's results as its output, rather than object code. The characteristics of compilers and interpreters are summarized in Table 7-6.

FORTRAN was the first high-level language to be widely accepted. Developed in 1954, it has subsequently been made available for

Table 7-6. Comparison of Compilers and Intrepreters

COMPILERS	INTERPRETERS
• Output is object program	• Output is executed result
• Execution time is faster	• Execution time is slower
• Development time is generally slower	• Development time is generally faster
• Flexible; may be interfaced with other languages	• Inflexible; may not usually be interfaced
• Programs may be saved in source or object form	• Programs may be saved in source form only
COMPILATION PROCESS	*INTERPRETIVE PROCESS*
1. *Compilation:* • Analysis and creation of table • Code generation • Optimization of code 2. *Assembly:* • Relocatable assembly into object code 3. *Linking:* • Linking of object into internal machine code	1. *Editing:* • On-line development and editing 2. *Run:* • Analysis and creation of table for one statement • Code generation for one statement • Absolute assembly of one statement into machine code • Execution of the operations for one statement and start of process for next logical statement

almost every computer ever produced. FORTRAN was the first programming language to have an industrywide standard. Hundreds of thousands of programs have been written in FORTRAN.

FORTRAN was originally intended for the scientific and engineering fields. Its primary feature is the easy conversion of complex formula and equations into FORTRAN statements. As an example:

$$S_p = \sqrt{\frac{P(10.7-P)}{N^2}}$$

would appear in FORTRAN as:

$$SP = SQRT\ (P* (10.7-P)\ /N**2)$$

Standard algebraic notation is expressed in a similar fashion within FORTRAN. (In fact, FORTRAN provides the opportunity to specify formats for input and output operations. This permits easy, readable output of the tables of numerical values, characteristic of scientific and engineering problems.)

For a number of reasons, most FORTRAN translators are compilers rather than interpreters:

1. Once developed, scientific and engineering programs are often run many times without changes;
2. Frequently, problems in FORTRAN are lengthy. Faster execution time with a compiler is an important advantage;
3. Relative to critical operations requiring high efficiency, assembly language subroutines may be assembled into object code. These are then linked together with the output of the FORTRAN compiler, forming a single machine code program.

To conclude, one of FORTRAN's most glaring deficiencies is the absence of an alphabetic variable type. This has eliminated FORTRAN from effective use in many nonnumerical applications, such as text processing, many business operations and other activities where letter manipulation is more critical than numbers manipulation.

However, FORTRAN has the advantage of a powerful subroutine capability. By using subroutines, the programmer can break a program into several smaller subroutines (or modules). The modular approach to programming speeds the program development process, as smaller modules are easier to code and correct. As such, marked reduction of program development costs is possible.

BASIC is an acronym for Beginner's All Purpose Symbolic Instruction Code. Developed at Dartmouth College in 1965, it is intended to be a very simple language to learn. It is used primarily to train beginning programmers.

BASIC is very similar to FORTRAN in the expression of mathematical equations.

Thus, the equation

$$c = \sqrt{a^2 + b^2}$$

becomes

Let C = SQR (A ↑ 2 + B ↑ 2)

Unlike FORTRAN, however, BASIC employs simpler input and output statements. No complicated format specifications are required. To print the value of the variable X at the user's terminal, for example, the programmer need only write:

PRINT X

Normally, BASIC translators are interpreters. This is so because the language's emphasis is on training, and the interpreter's error detection-correction feature is an important capability. However, BASIC compilers are appearing as former student users of BASIC enter the scientific, engineering, and business worlds. Some applications presently written in FORTRAN and COBOL, therefore, are also beginning to appear in extended versions of BASIC. BASIC is now particularly being used for business applications on several minicomputers where COBOL cannot be properly implemented.

BASIC has good capabilities for manipulating letters as data. Applications such as text processing and report generation, are easily written in BASIC. Conversely, BASIC has weak subroutine characteristics; modular programming is not easily accomplished.

Although FORTRAN and BASIC were designed to benefit different application areas, there is some overlap. Generally, however, some sacrifice in overall efficiency is necessary when these languages are used outside of their primary areas.

TYPES OF SOFTWARE

There is a variety of software packages essential to the operation of a computer. They can be segregated into three types: *System software, command software,* and *applications/EMCS front end software.*

System Software

System software is the basic software which allows the computer to function, and may be referred to as "firmware." It includes programs to handle I/O devices, to load command and applications software and to develop any other software necessary to upgrade the EMCS.

Some of these programs comprise a user-programmable capability and include editors, assembler, compiler relocatable linking loaders, debuggers and interpreters. User programmability is a desirable feature, but there are distinctions between various systems on this feature. Two basic categories of systems exist: Limited User Interface and Expanded User Interface. With each category there are systems that vary in degree.

Limited User Interface Systems are typified by the minicomputer-based systems. These allow user interface for monitoring and changing parameters only. Several variations on this theme exist with microprocessor-based systems as well. Essentially these EMCS employ library algorithms for control, and the user's function is limited to changing time and setpoint parameters. Programming beyond that level is done by the manufacturer's technicians.

Expanded User Interface Systems also vary in degree; however, as a rule programming requirements are greatly increased. Programming is done in an EMCS language, as noted, and usually conforms to an English language type of format. These systems typically require that the user or programmer write all of the control algorithms. In addition it may be necessary to develop sophisticated mathematical formulas for the more complex strategies discussed later, such as optimization. A number of issues must be carefully weighed in considering this type of system due to the level of support which must be provided. However, many users like these systems because of the flexibility they offer in developing control strategies. Chapter 17 on EMCS Program Management discusses the concept of software style in greater detail, along with some of the practical considerations.

Multi-User or Multi-Tasking Software

Multiprogramming allows two or more programs to be executed by the CPU concurrently. Each program runs in a separate area of memory (called a partition) and uses the CPU alternately. Only one program executes at any given instant, but while one program is waiting for an I/O device, another program can be executed. Figure 7-11 shows the increased CPU and I/O efficiency of multiprogramming, providing a much higher utilization of computer resources.

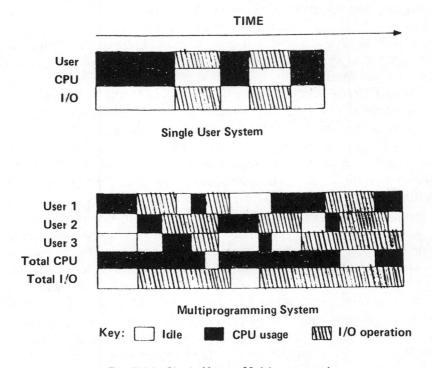

Fig. 7-11. Single User vs Multiprogramming

Operating Systems: Foreground/Background

A combination of on-line processing sometimes is necessary, especially in real-time control applications. The foreground is the memory area that holds the real-time programs; the area that holds noncritical batch-type programs is called the background. Foreground programs are time-critical tasks, such as process control. Further, time-critical tasks are identified by interaction with mechanisms such as remote sensors. Rather than human orientation, these devices are mechanically oriented.

Conversely, background programs are low-priority tasks. These run when the foreground program is not in use. Typically, background tasks include program development, report generation, maintenance, inventory reporting, etc., essentially, those tasks which are not time critical. In sum, background task runs until one of the following occurs.

1. The computer operator interrupts the background program;

2. a scheduled event occurs. Interrupts of this type can be triggered by the time of day, an elapsed time counter or a periodic schedule;

3. an operating system routine is called by the background program;

4. a hardware interrupt for a foreground task occurs.

Foreground/background systems, then, efficiently use an entire computer system. Real-time control occurs in the foreground; when no time-critical program occupies the foreground, a background program can produce payrolls, reports, mailing lists, etc.

Specified parts of a single-user program may take turns in main memory, by means of an overlaying technique. This technique is helpful when a program is physically larger than the available memory. When a program has subroutines, overlaying lets the programmer bring a specific subroutine into memory when it is needed. In other instances, memory locations can be used by other subroutines.

Command Software

Command software is a collection of specialized programs that allows the operator to interface with the EMC system and perform required actions such as changing parameters or monitoring a point.

As discussed under system software, the EMCS style will determine whether command software is employed. Command software ranges from menu-driven prompts for monitoring and programming, to a manufacturer-specific code which must be mastered to perform any function. For example, these programs may include command line interpreters, report generators, graphic software and control sequence software. The command line interpreter is a program that takes simplified English commands from an input keyboard and translates into machine code that can be executed by the EMC system.

Applications Software

Applications software is the set of programs which cause specified functions, such as demand control or duty-cycling to occur. Functions commonly performed by EMC systems are discussed in the following sections. Applications software also includes programs which are resident in a personal computer or remote terminal. These programs provide a wide variety of functions which are used in communications and program management.

This chapter has provided an overview of EMCS and computer technology. It has been the intention of this text to provide the reader with the basis for the discussion of EMCS hardware types which follows in Chapter 8. That discussion will focus on specific characteristics, and address key technology issues regarding each of these system types.

8

Level III EMCS Hardware

As noted in Chapter 7, Level III systems are divided into "hardware types" based upon the equipment in use: microprocessor- and mini-computer-based systems. In Chapter 8, these hardware types will be distinguished from one another under several headings. These include:

Computer Technology
Peripheral Devices
Field Equipment
Software
Miscellaneous Cost Issues
System Architecture

The primary focus of this chapter will be System Architecture, due to the deviation between the hardware types with regard to this issue. However, specific distinctions will be addressed under each heading which must be considered in system design. It is essential to secure the services of a qualified EMCS consultant in this process, because there are many variables which must be considered in specifying a hardware type.

Prior to addressing specific EMCS issues under the above headings, an expanded description of each hardware type is provided.

Minicomputer-Based Systems, as the name implies, are EMC systems built around a minicomputer. The processing-intensive nature of computerized control demands the use of these systems with many major EMCS installations. These sophisticated computer controllers are well suited to extensive controls projects, 500 points and up. Minicomputers often call for a dedicated control room, due to defined requirements for space temperature and humidity conditions. As will be discussed under specific headings, in the past these systems have been associated with sophisticated control options such as: analog outputs, HVAC optimization, fire and life safety features, etc.

Microprocessor-Based Systems, as the name implies, are EMC systems built around a microprocessor chip. As a result of advances in computer technology, these systems can typically be housed in self-contained enclosures. These devices can also operate in a wide range of ambient conditions. Developments in the computer industry have allowed for expanded system power, providing sophisticated control options, flexibility, and expanded point capacity with microprocessor-based systems. This has significantly increased the implementation of these systems. Today this type of system can meet most detailed EMCS specifications. The flexible operating temperature ranges have also allowed widespread implementation of local FIDs which employ microprocessors for distributed processing.

COMPUTER TECHNOLOGY

It is not the intent of this section to compare or contrast microprocessor and minicomputer technology. The discussion of computer technology in Chapter 7 is applicable to both of these hardware types, though distinctions exist with regard to the EMCS application. The focus of this section is to review some of those issues which should be carefully evaluated with a consultant or EMCS team.

As noted above, ambient operating temperature range is an important issue when designing an EMCS. It is also critical to consider the need for stand-alone local control with a piece of equipment, and space availability to accommodate a command center. If any of these issues are of concern, a microprocessor-based system may be better suited to the application. Distributed processing, as discussed in Chapter 7, is being established as the state-of-the-art in building control. The benefits provided by these systems are impressive, and microprocesser-based systems are essential to implementing this technology.

The issue of EMCS capability has also been discussed and must be considered. The minicomputer-based systems can offer large point capacity, ample room for expansion and in some cases superior processing and scan time for control. The extent of the application is important as well, because most minicomputer-based systems are approved by Underwriters Laboratories and other standards agencies for fire and life safety functions. If these features are required for the present scope of work or for future expansion, careful evaluation of each system should be conducted. Local codes must also be consulted to determine whether specific standards are to be met by the building system.

PERIPHERALS

The minicomputer-based system has changed relatively little in recent years, with regard to peripheral interface. The key component in today's system, the PC, has had limited introduction with this hardware type, and as a result remote communications remains hampered by proprietary hardware.

The real excitement in this area has been with the microprocessor-based systems. These systems tend to be self-contained in terms of peripheral input, output and auxiliary storage computer hardware. The developments have been with the remote computer interface, however these added features have not come without complication. As discussed in Chapter 7, the PC functions as a front end. It will also be covered below.

The key reason for mentioning this topic is to focus the prospective EMCS user on the issue of interface. Users must evaluate their needs for system interface, both local and remote, to ensure that the EMCS installed can provide the functionality required. The front end is one of the most essential components to a successful EMCS program, and the features available to the user in this area should be specified along with control functions.

FIELD EQUIPMENT

There is minimal variance between the field equipment employed by both hardware types. In effect, many of these devices are interchangeable. However, prospective users should again acquaint themselves with the equipment bid for their project, and determine whether it is satisfactory. At issue are such concerns as:

1. Sensor accuracy, reliability, and style;
2. the use of discrete versus multiplexed outputs; and
3. data transmission accuracy and reliability.

SOFTWARE

Software may be viewed in several contexts, but the focus in this section is on user interface. Some microprocessor-based systems, and nearly all minicomputer systems, restrict users to no more than monitoring and setpoint modification capability. This may be acceptable, however the long-term service and maintenance of EMCS software must be considered. If the user will have trouble justifying an ongoing service contract, or requires greater access to control algorithms, an alternate choice should be considered.

The issue of software is addressed in Chapter 17, under program management, and in Chapter 7. Again the differences between software styles are extensive, and it is important to evaluate each package with regard to the EMCS plan.

MISCELLANEOUS COST ISSUES

MAINTENANCE CONTRACTS

The recommendation to develop and support an ongoing EMCS maintenance program has been put forth several times in this text. It is essential to system performance to maintain the equipment. The distinction lies in the extent and cost of this support. As a rule minicomputer-based systems entail extensive and costly maintenance contracts, around 10% of the original system cost, year after year. One of the reasons for this is the software issue. Any reprogramming requires factory support.

To their detriment, microprocessor-based systems are often not maintained. However, it does appear that these systems are somewhat less susceptible to costly maintenance. In some cases these systems also lend themselves to in-house maintenance through skilled tradepersons.

FIRST COST

EMCS systems have traditionally been compared on a "cost per point" basis. Given the large number of features, functions and associated variables which must be considered with the EMCS, this is not always the best criterion for comparison. However, if first cost is a critical concern, the minicomputer-based system may not be the best choice.

Due to the extensive nature of the equipment, these systems typically have a high cost per point. Yet, the best criterion for system evaluation is the application. The system which is installed should be well matched to the application, with adequate room for anticipated expansion. Cost should be an issue, but not at the expense of function and ultimate system performance. In the final analysis cost is always an issue, but the services of a good consultant, and solid EMCS planning, are essential to weighing the issues of cost and functionality.

SYSTEM ARCHITECTURE

System Architecture refers to the general organization and structure of EMC system hardware and software. It indicates the number of types of components, and the manner in which they interrelate to perform the functions required to implement the applications involved.

System architecture is shown schematically. It does not refer to the actual physical placement of components in a building.

Figures 8-1 through 8-8 illustrate eight different types of system architecture. All but Figure 8-2 employ distributed processing, which is recommended. Figures 8-1 through 8-3 depict microprocessor-based systems, while Figures 8-4 through 8-8 involve minicomputer-based systems.

The systems illustrated increase in system capability, and it should be noted that none of these systems represent specific manufacturers, models, etc.

Figure 8-1 illustrates the architecture of the most basic microprocessor-based system. This is a self-contained EMC system. As with all the microprocessor-based systems shown, "self-contained" means

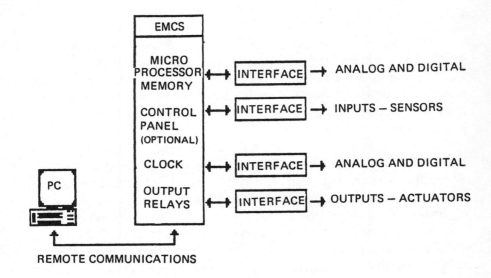

Figure 8-1. Microprocessor-Based EMC System Architecture

that all features necessary to maintain control are provided. These features include: CPU, memory (including 24-hour back-up for RAM), clock, communications and input/output (I/O) functions, which are all resident in one panel. This type of panel typically includes low voltage pilot relays, which allow the installer to interface directly with some loads. This system architecture depicts a relatively limited point capacity, usually between 8 and 32 input and output points combined. Inputs may be universal, meaning they can be defined as analog or digital, or specified as one type. Outputs are most typically digital only, though "pulse width modulation" is often provided to simulate analog outputs for modulating control. As shown, the system provides stand-alone control, and allows for local and remote communications. The communications option may employ dumb terminal or proprietary hardware and software.

Figure 8-2 shows a system which is basically the same as the one shown in Figure 8-1, except with more capacity. Note that this is the only system architecture shown that does not employ distributed processing. This is a common architecture in the industry, as it uses dumb FID panels to expand I/O. In this architecture, the primary, or

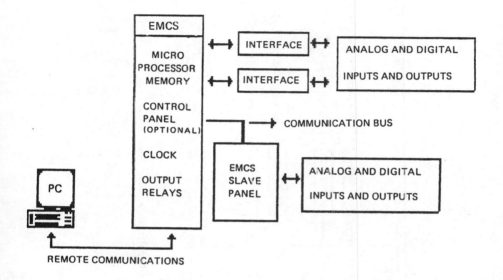

Figure 8-2. Microprocessor-Based EMC System Architecture (Alternate)

as it is typically called "master," unit contained all the same functions described under Figure 8-1. Therefore, a loss of communication between the master and subordinate or slave unit, or a power failure at the master unit, results in facility-wide loss of control. A variation on this architecture style which is also common has the master unit with no I/O capacity, thus increasing the system vulnerability. Neither of these styles are recommended due to the possibilities associated with an interruption in control caused by various problems with the master unit.

Figure 8-3 is again similar to the previous systems, but employs distributed processing to enhance capacity, reliability and expandability. This system depicts stand-alone panels which can communicate on a common bus to share data. A network form of communications is used to allow multiple smart panels to control loads based upon inputs directly connected, as well as from data acquired through other smart panels. The architecture of each smart panel may vary as well. A trend is underway in the industry to develop common communications protocols which would allow a VAV box controller to interface with a chiller control. In each case, the inputs

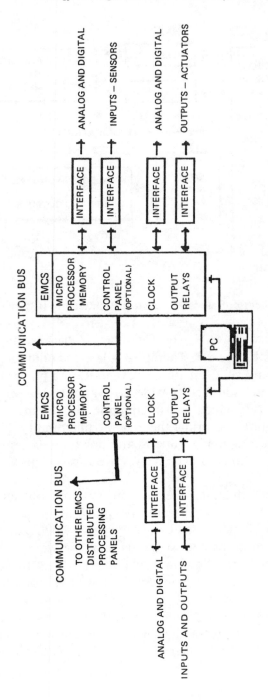

Figure 8-3. Microprocessor-Based EMC System Architecture (Alternate)

monitored and the loads controlled vary greatly, but the availability of common data enhances overall building performance. As with the other microprocessor-based systems, EMCS functions are self-contained and communications, either local or remote, are accomplished through external hardware. The most common vehicle for this function, as previously mentioned, is the PC.

The CPU in the minicomputer applications shown in Figures 8-4 and 8-5 is augmented with a microcomputer for control; however, this is not to be confused with a microprocessor application.

Figures 8-4 through 8-8 illustrate five different types of mini-computer-based system architecture, all employing distributive processing, which is recommended. With distributive processing, the system still is able to function even though the CCU—central control unit, a microcomputer or minicomputer—is down for any reason.

The systems again illustrate a hierarchy of system capability, reliability, and expandability. It should be noted that none of the systems represent specific types. In fact, the number of variations in systems is infinite, since a system can—and in most cases should—be designed to accommodate the specific needs of the facility involved.

Figure 8-4 illustrates the architecture of an EMC system designed for a relatively small building. The system's CCU is a microcomputer, which has relatively limited capacity. As such, the system is not designed with expansion in mind, although it can be expanded if need be. The system is intended primarily to handle basic functions and has limited optimization capability. The floppy disk shown serves to load and unload software; no bulk storage is provided. Each of the FIDs utilized in the system is a so-called smart FID; that is, a micro-computer-based controller, tying the data environment—all points monitored and/or controlled—to the communications network. The FID is a stand-alone device because it does not have to communicate with the CCU to perform its functions. (It communicates with the CCU on an exception basis only.) Each smart FID can be programmed to perform basic functions such as start/stop, duty-cycling, and alarm enunciation.

Figure 8-5 illustrates a system which is basically the same as the one shown in Figure 8-4, except more capacity, reliability and expandability are built in. These are provided through the use of a

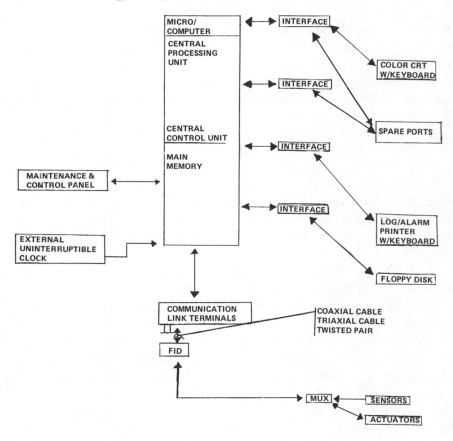

Figure 8-4. Minicomputer-Based EMC System Architecture

minicomputer CCU instead of a micrcomputer, and the addition of a CCC (Central Communications Controller). Like the CCU, the CCC is a general purpose digital minicomputer, with a central processing unit (CPU), memory, and an input/output capability. The CCU is the "brain" of the EMC system, supervising the FIDs through the communication network by executing programs stored in memory.

The CCC functions to unburden the CCU from communications functions. It is a stripped-down version of the CCU tied directly to the CCU and the communications network. The CCC handles the communication tasks of timing, switching, formatting and protocol.

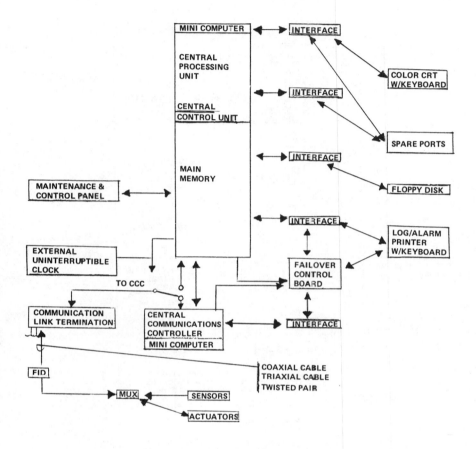

Figure 8-5. Minicomputer-Based EMC System Architecture (Alternate)

The CCC also provides reliability through redundancy. It can take over the functions of the CCU in the event that it is "down" for any reason.

The architecture shown in Figure 8-6 is identical to the one shown in Figure 8-5, except it has an auxiliary storage capability provided through a single dual-drive disk interfaced through a controller. The disk memory serves to increase the software capability of the system.

The system shown in Figure 8-7 is designed for very large facilities and complexes which need extensive optimization routines and have or intend to have an in-house minicomputer programming capability.

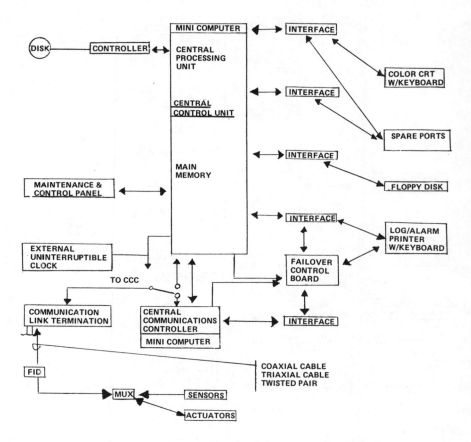

Figure 8-6. Minicomputer-Based EMC System Architecture (Alternate)

The basic difference between this system and the one shown in Figure 8-6 is the addition of a second disk memory to extend capabilities further; provision for a magnetic tape drive and controller for collection of historical data used later in development of more sophisticated optimization programs, and a black-and-white CRT with keyboard to enable a software development by one operator while another uses the color CRT for routine operating tasks.

The system shown in Figure 8-8 is the same as the one shown in Figure 8-7, except it has an additional CCC to provide increased reliability.

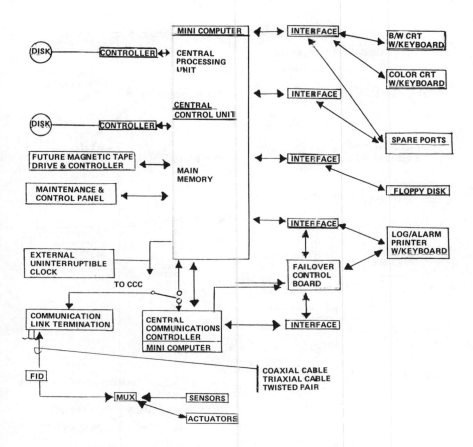

Figure 8-7. Minicomputer-Based EMC System Architecture (Alternate)

As indicated by these eight different systems, future requirements must be considered carefully. If it is likely that the system will be expanded, it is far less expensive to provide for expandability when designing the initial system than it is to add it later on. Naturally, these concerns should be addressed by the consultant who then can develop a system with life-cycle cost-effectiveness in mind.

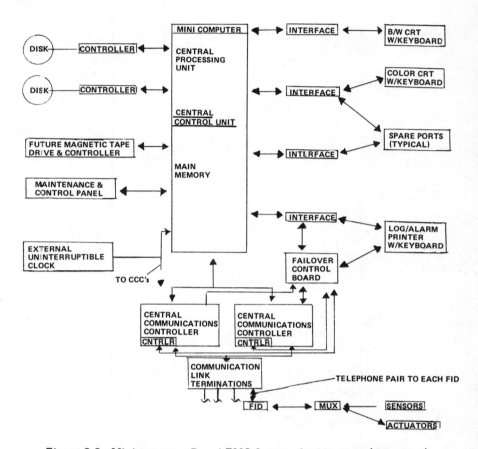

Figure 8-8. Minicomputer-Based EMC System Architecture (Alternate)

9

EMC System Functions and Applications

By definition, an EMC system, regardless of its level, performs its functions on systems. If it alters the operation of just one element of a system, the whole system still is affected. Accordingly, before any decision is made with regard to functions to be performed, the EMC system consultant must ascertain the types of functions which should be performed in light of characteristics unique to the building system involved, the purposes for which it is used, and the effect—if any—that application of an EMC system will have on system performance and output.

The following discussion relates primarily to Level III EMC systems. Nonetheless, the principles involved are applicable to all levels. Several other points deserve comment.

First, the functions which an EMC system can perform go far beyond those indicated below. Those discussed are limited primarily to energy conservation and systems optimization. Most EMC systems can also perform functions associated with building security, fire safety, and—in some cases—accounting and inventory maintenance.

Second, the functions indicated relate primarily to HVAC systems. Nonetheless, the same function could be applied to other systems. For example, remote start/stop can be applied to lighting systems just as easily as it is applied to fans and pumps. In this scenario, a function can achieve nonenergy related benefits in such areas as: enhanced building control or information management. An interesting case study outlining this type of application may be found in Chapter 23.

Third, no attempt has been made to differentiate between applications applying to critical and noncritical areas. For example, while night setback may be appropriate for an office, it would be completely inappropriate where given conditions must be maintained around-the-clock.

As in all other aspects of energy management, therefore, decisions must be made in light of the role of the facility; as desirable as energy conservation and optimization may be, it is even more important that the ability of the facilities to support the mission not be impaired in any way.

Generally speaking, functions performed can be segregated into three general categories: basic functions, optimizing functions, and operation functions, discussed as follows.

BASIC FUNCTIONS

Basic functions are those which are generally applicable to a broad range of systems and equipment. Some of these functions include the following.

Time Scheduled Operation

Time scheduled operation functions consist of starting and stopping a system based on time and type of day. (Type of day refers to weekdays, Saturdays, Sundays, holidays, or any other day which has an off-normal schedule or operation.) Time scheduled operation provides the greatest potential for energy conservation if systems are currently being operated during unoccupied hours. It also is the simplest function to install, maintain, and operate.

Duty Cycling

Duty cycling consists of shutting down a system for predetermined short periods of time during normal operating hours. Normally applicable to HVAC systems, the function is based on the theory that HVAC systems seldom operate at peak output. Accordingly, if a system is shut down for a short period of time, it still has enough capacity to overcome the slight temperature drift which may occur during the shutdown. Although the interruption does not reduce total cooling load, it does reduce the net auxiliary loads, such as fans and pumps. It also reduces outside air heating and cooling loads since the outside air intake damper is closed while an air handling unit is off.

Systems generally are cycled for a fixed period of time, such as

15 minutes of each hour of operation. The off period time length normally is increased during moderate seasons and reduced during peak seasons. Cycle fans utilized with direct expansion (DX) cooling coils or heat pumps, and cycle pumps providing chilled water flow through chillers should not be duty-cycled.

The subject of duty cycling has caused considerable controversy with regard to the damage which it may or may not cause to motors. The four primary concerns in determining whether or not duty cycling may cause damage are:

1. The number of stops and starts in a given period of time;

2. percentage of time for which the motor is off and on;

3. external inertia, or the amount of weight which the motor must move and at what speed; and

4. level of load, meaning the capacity at which the motor is operating when it is on.

Few, if any, problems will be experienced when the motor involved is relatively small; that is, approximately 50 HP or less. This experience is verified through studies which suggest that no damage will occur when the motor is 250 HP or less. In any event, if there is a question, especially when the motor drives a critical load, expert guidance should be obtained. In all cases, when the motor still is under a manufacturer's warranty or guarantee, contact the manufacturer to determine its suggestions for appropriate duty cycling and the effect which duty cycling may have on the warranty or guarantee.

Temperature Compensated Duty Cycling

The temperature compensated duty cycling function is essentially similar to duty cycling. Loads are assigned a maximum and minimum off time.

Each load has a related space temperature data point assigned to it. Each load serving an interior zone (cooling only) is cycled off for its assigned maximum period as temperature in the space drops to its assigned minimum value. Inversely, the interior load is cycled off for its assigned minimum period as space temperature rises to its assigned maximum value. For loads serving perimeter zones (summer-cooling, winter-heating), an outside air changeover sensor affects the algorithm used. With the outside air sensor determining mild or summer conditions, the "off" time calculation is the same as described for loads serving interior spaces. With the outside air sensor determining winter conditions, the "off" time calculation parameters are inverted, so minimum "off" times occur at minimum space temperatures. "Off" times vary proportionally throughout the assigned acceptable space temperature range.

Demand Control

Demand control function stops electrical loads to prevent a predetermined maximum electrical demand from being exceeded. Many complex schemes are used to accomplish this function. Generally speaking, most of these monitor the base electrical demand continuously and predict whether or not the preset limit will be exceeded. If it will be, certain scheduled electrical loads (secondary loads) are shut off. Additional secondary loads are turned off on a priority basis if the initial load shed action does not reduce the predicted demand enough to satisfy the function requirements.

When demand control is applied to water chillers, special considerations are necessary.

Water chillers are generally equipped with a manually adjustable control system which operates to control the chilled water supply or return temperature that the chiller operates against. An interface between the EMC system and this control current allows the EMC to reset the temperature setting in a load shedding situation, thereby reducing electric demand without shutting down the chiller completely. The method of accomplishing this function depends on the specific water chiller involved. In general, however, when the chiller is selected for load shedding, an analog reset signal is transmitted to the interface which then increases the chilled water temperature setpoint by a like amount. Chiller power demand is monitored by the chiller power meter and the chilled water temperature is reset upward until chiller power consumption is reduced to an operator setable minimum KWD setpoint. *Extreme caution must be exercised with application of this function.* Incorrect interface and control may cause the centrifugal refrigeration machine to operate in a surge condition, and may cause the reciprocating refrigeration machine to suffer from overheating of the hermetic motor or poor oil return, ultimately causing considerable damage to the equipment. Thus, the intent of this function is to leave intact the manufacturer's safety and operating controls and to reduce the power consumption through operation of the centrifugal vanes, or the reciprocating unloaders by resetting the chilled water setpoint. Reciprocating machines, like centrifugal machines, should be provided with manufacturer's recommended recycle timers to limit number of starts per hour.

Demand control also can be integrated with standby power generator operations. When electrical demand approaches a peak, this function starts the engine or turbine generators which feed electrical power into the building where they are located, or drive specific items of equipment such as well water pumps, thus reducing base electrical demand. Extreme caution must be utilized in using this function. Only the largest of generators should be considered since considerable investigation and expense may be necessary to perform any rewiring or reswitching needed for proper operation of this function.

Outside Air Temperature Cutoff

With an outside air temperature cutoff, heating systems except for direct-fired warm air systems smaller than 500,000 Btu per hour design input capacity, are provided with an outdoor temperature-sensing control which cuts off the heating system for all types of spaces when the outdoor temperature rises to within 5F of the indoor design temperature, or to a minimum of 40F.

Warm-Up/Night Cycle

The thermal load imposed by outside air used for ventilation may constitute a substantial percentage of the total heating and cooling requirements for a noncritical area. The warm-up/night cycle function controls the outside air dampers when the introduction of outside air would impose a thermal load and the area is unoccupied. This function would apply during warm-up or cool-down cycles prior to occupancy of the area and would also apply in certain facilities of electronic equipment, when the building is unoccupied. During those times, the outside air dampers would be closed.

Space Temperature Night Setback

The energy required to maintain space conditions during the unoccupied hours can be reduced by lowering the temperature set point for the space, depending on the climatic conditions. This function would apply only to noncritical areas that are not required to operate 24 hours per day. Normally, where applicable, this function would reduce the space temperature from the normal 68F winter inside design temperature to a 50F or 55F space temperature during the unoccupied hours.

OPTIMIZING FUNCTIONS

Optimizing functions are those which improve the operating efficiency of a given piece of equipment or system by monitoring performance and performing adjustment on a real-time basis. Typical optimizing functions include, but are not limited to the following.

Economizer Cycle

The utilization of an all-outside-air economizer cycle can be a cost-effective energy conservation measure, depending on the climatic conditions and the type of mechanical system. Where applicable, the cycle utilizes outside air to satisfy all or a portion of the building's cooling requirements when the temperature or the total heat content of the outside air is less than that of the return air from the space. Outside air is introduced through the mechanical system and relieved during this cycle instead of the normal recirculation system.

A dry bulb (db) economizer cycle is regulated by an outdoor air temperature sensor which discontinues compressor operation when outdoor temperature falls below the design supply temperature, normally 50F to 60F. Cooler outside air is then drawn into the system and used to reduce space temperature.

An enthalpy economizer cycle is similar to a dry bulb economizer cycle, except it measures total heat content (enthalpy) of air. Outdoor air is used for cooling when its enthalpy is less than the enthalpy of return air.

Choosing between a dry bulb or an enthalpy economizer cycle involves a decision which should be based on economics. If periods of high humidity seldom occur in your area, the additional expense of an enthalpy economizer will not be cost-justified.

Hot/Cold Deck Temperature Reset

HVAC systems such as dual duct and multizone utilize a parallel arrangement of heating and cooling surfaces (hot and cold deck surfaces) to provide simultaneous heating and cooling. Both heated and cooled air streams are mixed to satisfy thermal requirements in a given space. Without optimization controls, these systems are extremely wasteful, because the temperature settings of the hot and cold decks are fixed, usually at 90F and 55F, respectively. With temperature reset, the system selects the individual areas with the greatest heating and cooling requirements and adjusts hot and cold deck temperature accordingly, minimizing the inefficiency of the system by reducing the difference between hot and cold deck temperatures. Return air humidity may also be monitored for high limit.

Discharge Air Temperature Reset

The discharge air temperature reset function adjusts the cooling coil discharge temperature upward until the zone with greatest demand for cooling has closed its reheat coil valve. Return air humidity may be monitored for high limit.

Chilled Water Reset

The energy required to generate chilled water in a reciprocating or centrifugal electric-driven refrigeration machine is a function of a number of parameters including the temperature of the chilled water leaving the machine. Because the refrigerant suction temperature is a direct function of the leaving water temperature, the higher the two temperatures, the lower the energy input per ton of refrigeration. As a result, since chilled water temperatures are selected for peak design times, most chilled water temperatures can be elevated during most operating hours, unless strict humidity control is required. Depending on the operating hours, size of the equipment, and configuration of the system, energy can be saved by resetting the chilled water temperature, allowing it to rise to satisfy the greatest cooling requirements. Generally, this determination is made by the position of the chilled water valves on the various cooling systems. The positions of the control devices supplying the various cooling coils are monitored and the chilled water temperature is elevated until at least one control device is in the maximum position. Other control schemes may be necessary to satisfy different system configurations.

Outside Air Schedule Reset

Hot water heating systems are designed to supply system heating requirements at outdoor design temperatures. Depending on the specific system design, the hot water supply temperature can be reduced as heating requirements are reduced, usually in response to increased outdoor ambient temperatures. Where applicable, the capability to reduce the temperature of the supply water as a function of outdoor temperature will effect operating savings. To accomplish this function, the temperature controller for the hot water supply is reset on a predetermined schedule as a function of outdoor temperature.

Start/Stop Optimization

The optimized start/stop feature is an additional feature of the timed operation of mechanical systems described above. Mechanical systems serving areas that are not occupied 24-hours-a-day can be shut down during the unoccupied hours. Traditionally, the systems are restarted prior to occupancy to cool down or heat up the space in time for the first persons to arrive. Normally this function is performed on a fixed schedule independent of weather or space conditions. The optimized start/stop feature of the system automatically starts and stops the system to minimize the energy required to provide the desired environmental conditions during occupied hours. The function automatically evaluates the thermal inertia of the structure, the capacity of the system to either increase or reduce temperatures in the facility start-up and shut-down times, and weather conditions to accurately determine the minimum hours of operation of the HVAC system to satisfy the thermal requirements of the building.

Air Distribution Optimization

The air distribution optimization function involves control of zone dampers to stop the flow of conditioned air to nonessential areas during unoccupied periods. Dampers which in most cases will have to be installed, are positioned automatically by an optimum start/stop program. Supply fan dampers and associated exhaust fans are positioned by static pressure sensors.

Chiller Plant Optimization

Chiller efficiency usually decreases with chiller load below 50%. When a multiple chiller installation is involved, therefore, it generally is desirable to load each chiller as much as possible. In other words, it generally would be better to have two chillers loaded equally, as opposed to having one loaded to 100% and another to 30%. To achieve optimization, it is necessary to control loading.

Chiller loading can be controlled by resetting the supply and return water temperatures sensed at the unit. The temperature reset scheme should be selected in light of the specific installation involved, since there are many different variations. Some multiple chillers are connected in series, others in parallel, etc. Chillers with constant water flow can be controlled by return water temperature. Variable water flow must be controlled by supply water temperature.

A multiple chiller loading optimization function is one that controls temperature resetting in a manner to minimize energy consumption without affecting comfort conditions in any space served by the system. This normally is accomplished by measuring flow through the evaporator and condenser of each chiller and the reclaim coils. Differential temperature across evaporator sections, condenser sections, and reclaim sections also are monitored. This enables the computer to calculate the total energy being produced by each machine. Total electrical usage is monitored continuously thus enabling instantaneous efficiency calculations. When the temperatures are monitored the minimum energy requirement for the system can be calculated. The computer then will cause the most efficient chiller or combination to be operated.

Boiler Plant Optimization

Several techniques are available to accomplish the boiler plant optimization function, depending upon factors unique to the facility involved.

In one approach, electronic probes can be used to continually monitor the amount of O_2 (oxygen) that passes through the boiler plant stack. Data collected permits the CPU to determine heat losses associated with that stack. If the O_2 level is outside optimum limits, the CPU can initiate control to regulate air flow through the boiler and stack at constant firing rates, to bring the O_2 level within proper operating limits and retain maximum combustion efficiency.

Another method of boiler plant optimization involves decreasing the fuel flow to the boiler as load on the boiler decreases. During winter, higher outdoor air temperatures reduce building heat requirements, so reducing the amount of heat the boiler has to produce. Boiler outlet water temperature can be reduced by mixing the outlet water with boiler inlet water, to create a mixture determined by the outside air temperature and the boiler water temperature difference. This unloads the boiler and decreases fuel consumption. As the boiler unloads, flue gas temperature decreases, but still remains high enough to not precipitate corrosive substances inside the boiler stack. It is recommended that temperature sensors be placed at the top of the stack to warn if flue gas temperatures become too low. Depending upon the sulfur content of the fuel being burned, the flue gas temperature should be maintained above 250F to 300F. Also, automatic flue dampers should close when burners are off to prevent cooling of boilers and loss of heat.

Secondary Loop Chilled/Hot Water Optimization

There usually are many secondary loops serving similar coils, each tied into the primary loop main distribution system. The primary loop feeds the secondary loop with supply water from a central source and conveys the return water back to the central cooling or heating machinery, as shown in Figure 9-1.

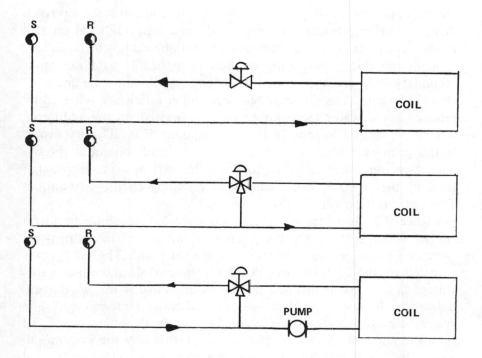

Fig. 9-1. Typical Secondary Water Applications

As can be seen, the control valve adjusts coil capacity to meet load by modulating coil flow, or by changing the temperature of water to the coil. Whenever the valve lowers coil output, it indicates that full capacity of the supply water is not needed. Accordingly, secondary loop optimization logic is based on the premise that if no coils need full supply water capacity, supply water temperature can be adjusted to save energy, by raising chilled water temperature and lowering hot water temperature. This action also saves energy by reducing pipe heat gains and losses.

Dehumidification must be considered when chilled water is affected. The highest tolerable chilled water temperature must not exceed the cooling coil control set point for dehumidification.

The optimization set-up involved requires stable local controls to ensure that final control elements of the distribution system (valves, dampers, mixing boxes) are measured accurately. If local control loops are not stable, this optimization technique cannot be used.

Although this optimization method is applicable to a variety of secondary loop arrangements, the system's layout must be analyzed closely. In some cases, energy saving in chiller efficiency achieved by raising chilled water temperature may be wasted on the additional pumping required to provide the same amount of total heat transfer. If the primary hydraulic loop has pressure head control at the by-pass, however, increased flow to the coils only means a decrease of flow in the by-pass, so no additional pumping energy is required. This optimization scheme then is applicable.

Figure 9-2 illustrates a typical secondary loop optimization of a chilled water system. Optimization is achieved by monitoring all valve end switches and adjustment of the set point. The set point is adjusted by the CPA (control point adjustment) until contact status voltage is lost. When this occurs, at least one valve is near wide open, indicating the need for "decreasing chilled water temperature." The system then decreases the set point T1 by a small value every fixed time interval until the contact is made. In this way the system balances itself dynamically to maintain a minimally sufficient chilled water temperature at all times.

End switches of the valves can be wired individually or in series. Individual wiring enables the CPU to know where coil is demanding more cooling, so a proper weighing factor can be used in the decision logic. A valve stroke transmitter can be used not only to indicate the near open position but also the actual control position of the valve. More sophistication may be incorporated in the software to determine the actions to be taken. Among all these instrumentation techniques, the serial wiring of the valve end switch is the most economical and least sophisticated method.

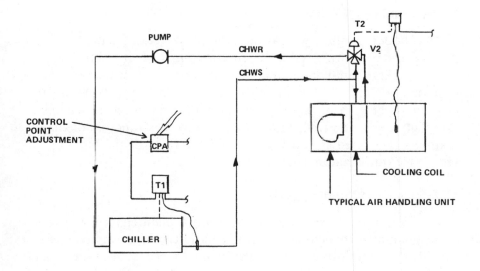

Fig. 9-2. Typical Secondary Loop Chilled Water Optimization

OPERATIONAL FUNCTIONS

Operation functions are primarily those which are associated with monitoring and alarm, as follows.

Chiller Profile Generation

Through a chiller profile generation function, the EMC system develops a printed record of the operating status of the chiller equipment for a given set of conditions, such as load ambient conditions and time of day. The program accumulates chiller profile historical data required to evaluate efficiency and cost of operation. These data include such parameters as chiller demand in kilowatts, return and supply chilled water temperatures, entering and leaving condenser water temperatures, chiller water flow in GPM, cooling load in tons and energy consumption in kWh/ton cooling.

Boiler Profile Generation

The boiler profile generation records operating data of boiler plant equipment for a given set of conditions, such as load, energy output, energy input and ambient conditions, to determine operating costs and to accumulate historical data for the plant. When a sufficient number of profiles of the plant have been obtained, curves of the plant can be derived, plotting efficiency as a function of the operating characteristics. Curves thus derived become the basis of reference for optimization routines.

Maintenance of Building Systems

An EMC system cannot actually perform maintenance, but it can be programmed (depending on the specific type system involved) to provide daily printouts of regular maintenance procedures which must be undertaken, together with as-needed notification of unscheduled maintenance procedures which may be required. There are several ways in which the maintenance scheduling functions can be handled, as follows.

Calendar Time Scheduling: The calendar time method obviously is the easiest and most direct way to schedule maintenance. For example, every 20 days maintenance men could be instructed to change all air handling unit filters, or once-a-month instructions can be given to grease and lubricate certain equipment.

Machine Running Time Scheduling: Since all machinery does not run on a regular schedule, calendar time scheduling for many items is not sufficient. Accordingly, the central processing unit can be set to accumulate running time of certain equipment and, after a predetermined number of running hours for each item, print out maintenance instructions.

Efficiency Scheduling: In certain cases the amount of time equipment runs will not be an accurate indicator of its operating efficiency or need for maintenance. Since the CPU can calculate equipment operating efficiency on the basis of raw data inputs, it can be programmed to provide maintenance instructions when efficiency deteriorates to a certain predetermined level.

Early Warning Monitor: The system can be designed to provide early warnings of impending equipment failure. As one example, bearings of certain pieces of equipment can be monitored for vibration and/or temperature. Should the vibration or temperature level increase to a certain predetermined level which indicates a problem, a panel light, readout, etc., would indicate an immediate, unscheduled maintenance problem. It also would be feasible for the computer to stop the particular piece of equipment involved—or the entire system of which it is a component, if necessary—until the needed maintenance or repair is performed.

Maintenance Cost Management Program: A maintenance cost management program stores maintenance costs and maintenance labor for individual units, fans, motors, pumps, etc.

Maintenance Personnel Scheduling: A program can be developed to provide daily or weekly maintenance personnel scheduling, identifying what has to be done, who should be doing it, etc.

Remote Communications

Although communications may not be specifically defined as system function, it is perhaps one of the single most essential system features. Communications is the key automation feature for remote programming, optimizing equipment and EMC performance and reducing service and maintenance costs. For more on this topic see Chapter 11.

Trouble Diagnosis

By monitoring certain parameters of the mechanical/electrical systems, diagnosis of reported problems with mechanical and electrical systems can be performed at the central console location. Some of the parameters that might be monitored for the purposes of trouble diagnosis include hot and cold deck temperatures with high and low limits, leaving chilled water temperatures and hot water temperatures with high and low limits, differential pressure switches indicating fan and pump operation or excessive pressure losses through filters or condensor tube bundles, space temperatures, and improper humidifier operation. Typical trouble diagnosis applications include those for:

1. *Refrigeration compressors* with capacity greater than 20 tons, provide separate status contact for each sensor listed; with capacity greater than 5 tons but less than or equal to 20 tons, provide single status contact for all sensors listed.

2. *Air handling filter banks* with (1) media depth greater than 3 inches, (2) media efficiency greater than 40% ASHRAE Atmospheric Dust or 90% ASHRAE Weight Arrestance, or (3) capacity greater than 6000 CFM; provide a filter differential pressure status sensor.

3. *Condensor tube bundles* operating with water from "open" circuit cooling towers; provide differential pressure status sensor contact to indicate excessive pressure differential, such as that caused by tube fouling.

4. *Humidifier installations;* provide alarm status contact to indicate simultaneous operation of the humidifier and cooling coil or economizer.

Safety Alarms

Many items of mechanical equipment are provided with various types of alarms for both personnel and equipment protection. These alarms such as high and low water for boilers, gas pressure alarms, and various temperature and pressure alarms on refrigeration machines, are typical of the types of functions that can be monitored. Monitoring of such alarms provides the console operator with information regarding the failure of equipment or the development of potential problems with the system operation.

Intercom

A function of the system that can be beneficial from a maintenance standpoint is the capability of communicating between the operator's console, the field interface devices or other remote intercom station locations. The system can be used for intercommunication between maintenance personnel and console operators for a checkout of the overall system, and for monitoring of start-up of equipment.

Metering

This function is to accept input signals in the form of binary contact closures and analog values from building and/or equipment energy meters to be installed in new buildings. The FID maintains a continuously updated cumulative meter reading in FID memory, calculates, using stored meter constants, a weighted 5-minute average consumption value in engineering units, stores these values for up to 30 minutes in FID memory, and transfers these 5-minute weighted averages to the EMC system CCU so that a building energy file can be established and maintained. A building energy file should be established for each meter installed, and the file should contain a file record for each elapsed calendar day. The file record should contain, in addition to file identifying information, the 5-minute weighted average consumption in 288 fields representing each 5-minute period in the day. This information can be utilized to examine building operating strategies and to verify the energy savings of the many different EMC system functions.

Typical metering functions include:

1. Building electrical consumption (KWH) and demand (KWD) for each building, project, or facility covered.

2. Cooling electrical consumption (KWH) and demand (KWD) for each electric driven refrigeration compressor with capacity greater than 40 tons.

3. Cooling steam consumption (pounds-hr) for each absorption refrigeration machine with capacity greater than 50 tons.

4. Cooling load (ton-hours) for each chilled water system with capacity greater than 50 tons. Wherever possible, locate water flow meter and supply and return water temperature sensors so as to allow metering of both cooling load and heating load from or to the building.

5. Heating load (Btu-hours) for each heating water system with capacity greater than 500,000 Btu per hour. Whenever possible, locate water flow meter and supply and return water temperature sensors so as to allow metering of both cooling load and heating load from or to the building. Utilize water meter-

ing for hot water heating systems fired by oil, gas, coal, or refuse, and for systems exchanging heat from steam, geothermal, or solar sources.

6. Heating steam consumption (pounds-hr) for each heating system with capacity greater than 500,000 Btu per hour utilizing steam directly in convectors, radiators, unit heaters, and AHU or H&V coils.

7. Heating fuel oil consumption (gallon-hour) for each fuel oil fired steam boiler or direct fired furnace with capacity greater than 1,000,000 Btu per hour input capacity.

8. Heating fuel gas consumption (Btu-hour) for each gas fired steam boiler or direct fired furnace with capacity greater than 1,000,000 Btu per hour input capacity.

9. Solar domestic water system (Btu-hour) for each solar heated domestic water heater with collector area greater than 1000 square feet.

Information Management

Employing the combination of computer technology and EMC-generated data provides the user with a powerful information tool. The data available as a result of the functions outlined in this chapter can help to insure system performance, and contribute to informed decision making. This is particularly true where remote communications allows monitoring and programming of the system. Managers find that current data on facility operations can be used to improve comfort, develop a more effective maintenance program and test the limits of their imagination for new ways to put this information to work. For more information on this feature, see Chapter 17 on program management.

OTHER FUNCTIONS

As mentioned, EMC systems can be programmed to perform numerous other functions not associated with HVAC systems. Some of the most commonly applied functions include the following.

Centralized lighting control can be provided through a flexible lighting control program that reduces consumption of electricity. Lighting in all areas could be analyzed to result in recommendations for reduced light levels and lighting shutdowns, where practical.

A *water and sewage usage monitoring system* can aid in reducing water consumption and sewage back-up effects, and as a management information tool. The system could monitor and record water and sewage flows and alarm operators of an undesirable condition.

An *access control program* that prevents unauthorized use of entrances and exits. Doors can be equipped with electric door locks controlled from the central guard station. An intercom station at each door would allow the guard to communicate with a person desiring entry into the building. If the guard determines that the person is authorized to enter, he could unlock the door by operating a switch on the guard console. For more positive identification, a closed circuit television camera could be located at each door so that the guard could observe the person desiring entry on a television monitor at his console.

A *smoke and fire management system* for optimum life safety, and to minimize property damage from fire and smoke. The system entails tying together all of building operating systems, including mechanical, electrical and communications systems so they work together for pressurization to control smoke, protect life and help fight fires. The existing fire alarm system can be interfaced and arranged to automatically control the building's heating, ventilating and air conditioning systems to help contain and prevent the spread of smoke in the event of a fire.

SYSTEM AND EQUIPMENT APPLICATIONS

There are a variety of systems and equipment used in buildings. Major HVAC system components which are most suitable for integration into an EMC system are shown in Table 9-1. It may not be cost-effective to connect systems and equipment which are of a capacity greater than that indicated in the table.

Table 9-1. HVAC System and Equipment Applications

HVAC systems and equipment to be integrated into Level III EMC systems, should have capacities greater than those listed below.

SYSTEM APPLICATION	CAPACITY
Single Zone Air Handler	2,000 CFM
Single Zone Split System	2,000 CFM
Terminal Reheat Air Handler	2,000 CFM
Multizone Air Handler	2,000 CFM
Two Pipe Fan Coil System	50 Tons
Heating and Ventilating Unit	3,000 CFM
Hot Water Convector Heating System	500,000 Btuh
Steam Convector Heating System	500,000 Btuh
Steam Unit Heater System	500,000 Btuh
Chiller, Air Cooled	15 Tons
Chiller, Water Cooled	15 Tons
DX Unit, Air Cooled	15 Tons
DX Unit, Water Cooled	15 Tons
Hot Water Boiler	500,000 Btuh
Steam Boiler	500,000 Btuh
Direct Fired Furnace	500,000 Btuh
Hot Water Convertor	500,000 Btuh

Tables 9-2 through 9-18 comprise matrices which indicate the functions applicable to the systems involved, and the input and output involved.

A separate matrix has been prepared for each major system. Each is easy to use. For example, Table 9-2 relates to a single zone air handler. If time scheduled operation of the system is desired, reading left to right indicates that only one input is required (a fan flow status sensor), and that the output is start/stop control interface.

	INPUT				OUTPUT	
Table 9-2. **Single Zone Air Handler**	Fan Flow Status Sensor	Space Temperature	Filter Differential Pressure Status	Humidifier Status	Start/Stop Control Interface	Local Control Interruption Interface
Time Scheduled Operation	X				X	
Space Night Setback	X	X			X	
Start/Stop Optimization	X	X			X	
Duty Cycling	X	X			X	
Demand Limiting Start/Stop	X	X			X	
Warm-Up/Night Cycle	X					X
Maintenance Run Time Report	X					
Trouble Diagnosis	X		X	X		
Table 9-3. **Heating and Ventilating Unit**						
Time Scheduled Operation	X				X	
Space Night Setback	X				X	
Start/Stop Optimization	X	X			X	
Duty Cycling	X	X			X	
Demand Limiting Start/Stop	X				X	
Warm-Up/Night Cycle	X					X
Maintenance Run Time Report	X					
Trouble Diagnosis	X		X	X		

Table 9-4. Single Zone Split System	INPUT				OUTPUT	
	Fan Flow Status Sensor	Space Temperature	Humidifier Status	Filter Differential Pressure Status	Start/Stop Control Interface	Local Control Interruption Interface
Time Scheduled Operation	X				X	
Space Night Setback	X	X			X	
Start/Stop Optimization	X	X			X	
Warm-Up/Night Cycle	X					X
Maintenance Run Time Report	X					
Trouble Diagnosis	X		X	X		

Table 9-5. Hot Water Convector Heating System	Pump Flow Status Sensor	Space Temperature			Start/Stop Control Interface	
Outside Air Temp. Cutoff	X				X	
Space Night Setback	X	X			X	
Start/Stop Optimization	X	X			X	
Maintenance Run Time Report	X					

Table 9-6. Terminal Reheat Air Handler

		Time Scheduled Operation	Space Night Setback	Start/Stop Optimization	Duty Cycling	Demand Limiting Start/Stop	Warm-Up/Night Cycle	Cold Deck Temperature Reset	Hot Deck Temperature Reset	Trouble Diagnosis	Maintenance Run Time Report
OUTPUT	Controller Reset Interface							×	×		
	Local Control Interruption Interface						×				
	Start/Stop Control Interface	×	×	×	×	×					
INPUT	Humidifier Status									×	
	Filter Differential Pressure Status									×	
	Greatest Heating Demand Sig.								×		
	Greatest Cooling Demand Sig.							×			
	Supply Air Temperature							×	×		
	Space Temperature		×	×	×	×					
	Fan Flow Status Sensor	×	×	×	×	×	×	×	×	×	×

Table 9-7. Multizone Air Handler

		Time Scheduled Operation	Space Night Setback	Start/Stop Optimization	Duty Cycling	Demand Limiting Start/Stop	Warm-Up/Night Cycle	Hot/Cold Deck Reset	Maintenance Run Time Report	Trouble Diagnosis
OUTPUT	Cold Deck Controller Reset Interface							X		
	Hot Deck Controller Reset Interface							X		
	Local Control Interruption Interface						X			
	Start/Stop Control Interface	X	X	X	X	X				
INPUT	Filter Differential Pressure Status									X
	Greatest Cooling Demand Sig.							X		
	Greatest Heating Demand Sig.							X		
	Cold Deck Supply Air Temp.							X		
	Hot Deck Supply Air Temp.							X		
	Space Temperature		X	X	X	X				
	Fan Flow Status Sensor	X	X	X	X	X	X	X	X	

Table 9-8. Two-Pipe Fan Coil System	INPUT					OUTPUT
	Pump Flow Status Sensor	Space Temperature (Several Typical Spaces)	Dual Temperature Water Flow Quantity	Dual Temperature Water Supply Temperature	Dual Temperature Water Return Temperature	Start/Stop Control Interface
Time Scheduled Operation	X					X
Outside Air Temp. Cutoff	X					X
Space Night Setback	X	X				X
Start/Stop Optimization	X	X				X
Duty Cycling	X	X				X
Demand Limiting Start/Stop	X	X				X
Maintenance Run Time Report	X					
Metering	X		X	X	X	

Table 9-9. Hot Water Convertor	INPUT					OUTPUT
	Pump Flow Status Sensor	Heating Water Flow Quantity	Heating Water Supply Temperature	Heating Water Return Temperature		Start/Stop Control Interface
Outside Air Temp. Cutoff	X					X
Metering	X	X	X	X		

Table 9-10. Steam Convector Heating System	INPUT				OUTPUT	
	Steam Flow Status Sensor	Space Temperature	Steam Flow Quantity	Steam Pressure Sensor	Local Control Interrupting Interface or Start/Stop Control	Start/Stop Control Interface
Time Scheduled Operation	X				X	
Outside Air Temp. Cutoff	X					X
Space Night Setback	X	X			X	
Start/Stop Optimization	X	X			X	
Metering	X		X	X		
Table 9-11. Steam Unit Heater System						
Time Scheduled Operation	X				X	
Outside Air Temp. Cutoff	X					X
Space Night Setback	X	X				X
Start/Stop Optimization	X	X			X or	X
Metering	X		X	X		

Table 9-12. Chiller, Air-Cooled

		Time Scheduled Operation	Demand Limiting, Chilled	Water Temperature Reset	Chilled Water Reset	Start Time Optimization	Maintenance Run Time Report	Trouble Diagnosis	Metering
OUTPUT	Controller Reset Interface					X			
	Chilled Water Reset Interface		X						
	Start/Stop Control Interface	X				X			
INPUT	Compressor Electric Consumption (KWH) and Demand (KWD)								X
	Dual Temperature Water Return Temperature								X
	Dual Temperature Water Supply Temperature								X
	Dual Temperature Water Flow Quantity								X
	Cooler Low Temperature Safety Status							X	
	Compressor Low Pressure Safety Status							X	
	Compressor High Pressure Safety Status							X	
	Low Oil Pressure Safety Status							X	
	Condenser Water Pump Status Sensor						X		
	Chilled Water Pump Flow Status Sensor						X		
	Chilled Water Supply Temp.				X				
	Chilled Water Return Temp.				X				
	Compressor Electrical Demand Meter (KWD)		X						
	Pump Flow Status Sensor	X	X		X		X	X	X

Table 9-13. Chiller, Water Cooled

		Time Scheduled Operation	Demand Limiting, Chilled Water Temperature Reset	Chilled Water Reset	Start Time Optimization	Maintenance Run Time Report	Trouble Diagnosis	Metering	Condensor Water Reset
OUTPUT	Controller Reset Interface			×					×
OUTPUT	Chilled Water Temperature Reset		×						
OUTPUT	Start/Stop Control Interface	×			×				
INPUT	Condensor Differential—Pressure Sensor Status						×		
INPUT	Condensor Water Supply Temp.								×
INPUT	Condensor Water Return Temp.								×
INPUT	Compressor Electric Consumption (KWH) & Demand (KWD)							×	
INPUT	Dual Temperature Water Return Temp.							×	
INPUT	Dual Temperature Water Supply Temp.							×	
INPUT	Dual Temperature Water Flow Quantity							×	
INPUT	Cooler Low Temperature Safety Status						×		
INPUT	Compressor Low Pressure Safety Status						×		
INPUT	Compressor High Pressure Safety Status						×		
INPUT	Low Oil Pressure Safety Status						×		
INPUT	Condensor Water Pump Status Sensor	×	×		×	×	×		×
INPUT	Chilled Water Pump Flow Status Sensor	×	×	×	×	×	×		
INPUT	Chilled Water Supply Temp.			×					
INPUT	Chilled Water Return Temp.			×					
INPUT	Compressor Electrical Demand Meter (KWD)		×						
INPUT	Pump Flow Status Sensor							×	

Table 9-14. DX Unit, Water Cooled

	INPUT										OUTPUT	
	Pump Flow Status Sensor	Condensor Water Pump Flow Status Sensor	Condensor Water Return Temp.	Condensor Water Supply Temp.	Chiller Tube Bundle Pressure—Differential Status Sensor	Low Oil Pressure Safety Status	Compressor High Pressure Safety Status	Compressor Low Pressure Safety Status	Condensor Low Temperature Safety Status	Compressor Electrical Consumption (KWH) & Demand (KWD)	Start/Stop Control Interface	Controller Reset Interface
Time Scheduled Operation	X										X	
Start Time Optimization	X										X	
Condensor Water Reset		X	X	X								X
Maintenance Run Time Report	X											
Trouble Diagnosis	X					X	X	X	X	X		
Metering	X									X		

Table 9-15. DX Unit, Air Cooled

	INPUT						OUTPUT
	Status Contact Sensor	On/Off Status Contact Sensor	Low Oil Pressure Safety Status	Compressor High Pressure Safety Status	Compressor Low Pressure Safety Status	Compressor Electrical Consumption (KWH) & Demand (KWD)	Start/Stop Control Interface
Time Scheduled Operation	X						X
Start Time Optimization	X						X
Maintenance Run Time Report	X						
Trouble Diagnosis		X	X	X	X		
Metering		X				X	

Table 9-16. Hot Water Boiler	INPUT							OUTPUT
	Pump Flow Status Sensor	Low Water Safety Status	Flame Failure Safety Status	High Temp. Safety Status	Dual Temp. Water Supply Temp.	Dual Temp. Water Flow Quantity	Dual Temp. Water Return Temp.	Start/Stop Control Interface
Outside Air Temp. Cutoff	X							X
Maintenance Run Time Report	X							
Trouble Diagnosis	X	X	X	X				
Metering	X				X	X	X	

Table 9-17. Steam Boiler	INPUT							OUTPUT
	Steam Flow Status Sensor	Steam Supply Pressure	Low Water Safety Status	Low Low Water Safety Status	Flame Failure Safety Status	High Pressure Safety Status	Fuel Flow Quantity	Start/Stop Control Interface
Outside Air Temp. Cutoff	X							X
Trouble Diagnosis		X	X	X	X	X		
Metering	X						X	

	INPUT					OUTPUT	
Table 9-18. Direct Fired Furnace	Fan Flow Status Sensor	Space Temperature	Flame Failure Cutoff Safety	High Temp. Cutoff Safety	Fuel Flow Quantity	Start/Stop Control Interface	Local Control Interruption Interface
Time Scheduled Operation	X					X	
Outside Air Temp. Cutoff	X					X	
Space Night Setback	X	X				X	
Start/Stop Optimization	X	X				X	
Warm-Up/Night Cycle	X						X
Maintenance Run Time Report	X						
Trouble Diagnosis	X		X	X			
Metering	X				X		

10

Direct Digital Control

Harris Bynum,
Roger Henderson
Regional Automation Specialists
Honeywell, Inc.

The topic of Direct Digital Control (DDC) is described here as a distinct concept. It is important to distinguish between the Energy Management and Control System (EMCS) which has been described as a computer, and DDC which is a method of processing and control. DDC control strategies are employed by many EMC systems today and they serve to enhance the control sophistication of these systems.

This chapter is intended to introduce a user or building HVAC system designer to DDC. DDC is different from previous applications of controls and computers within buildings. DDC is over 25 years old as a technology, but its introduction and frequent acceptance into the HVAC control arena, is much more recent. The technology evolutions that brought DDC to this economic threshold are continuing to improve the DDC product. Thin film laser trimmed platinum temperature sensors and new forms of solid state computer memory promise an even more exciting future as these innovations get blended into DDC products.

As with the introduction of DDC to the process control marketplace a few years ago, the introduction of DDC to commercial applications will sometimes be a painful and costly learning experience. With DDC, the effects of a component failure will be different, the effects of a failure of electrical power will be different, and the user's response to failures will be different. It is hoped that this introduction will assist users and designers in a more responsible and painless entry into this digital application.

This chapter will explore DDC under the following headings:

Definitions
Benefits
Adaptive Control

DEFINITIONS

Prior to defining DDC and discussing this technology in great detail, it is important to address two specific areas: DDC evaluation and DDC proximity.

In preceding chapters the topics of EMCS technology, hardware, software and applications have been discussed. Some confusion may arise from the start in discussing DDC, because it is not a component of an EMC system which can be evaluated like those mentioned above. DDC is primarily a function of software, however control strategies with this level of sophistication are typically transparent to the user. As noted in Chapter 7, software styles vary dramatically throughout the industry. But it is not typically possible to modify DDC control strategies. The feature which most users find effective is to modify parameters which are employed in the DDC control strategy, and in so doing effect changes to the nature of control.

As DDC is defined and discussed here, it will be evident that users would not normally want to modify the control algorithm itself. Rather, changes to the parameters will meet their needs. The results of this process are benefits, such as enhanced comfort and efficiency, which will be discussed later in this chapter. The decision that remains to be made by users is whether DDC is necessary, and the industry in general has determined that it is desirable. Therefore, the burden on users and their consultants is to insure that a reputable control manufacturer has effectively programmed this aspect of the software. It is hoped that this chapter will provide users with some tools to aid in making that determination.

The other issue of concern in discussing DDC is proximity. In true form, DDC control is interfaced directly at the device to be controlled via electronic apparatus. This method provides the optimum in accuracy and reliability due to the superior quality of electronic equipment. However, this form of interface is expensive, as electronic components have tended to be higher in cost. Therefore, many DDC controls have been interfaced in a hybrid, or supervisory, manner using a combination of electronic, electromechanical, and pneumatic controls. This method will be discussed later in the chapter. These systems function and provide benefits over many conventional controls systems. However, they may not meet the building owner's needs. As a result it is important that the EMCS consultant and the planning team be clear on the method of control interface, and that this is acceptable to the building owner.

From this point forward the discussion pertains to DDC as a control strategy, and not to the EMCS or the equipment that it controls. In order to proceed, the following terms must be somewhat understood.

1. *Closed Loop Control*—Pertaining to a system with feedback type control, such that the output is used to modify the input.

Thermostats are usually used in closed loop applications. An exception would be where an outdoor thermostat is used to position a hot water zone valve.

2. *Direct Digital Control*—A control loop in which a digital controller periodically updates the process as a function of a set of measured control variables and a given set of control algorithms.

As with most definitions of subjects as broad as DDC, this one is meaningless. That is, DDC could be a simple time clock, or a complete and total control system. For the purpose of this chapter, DDC is considered to be a total control system in which a digital computer monitors raw digital and analog sensor data, and its internal clock; and controls a "system" via analog and digital outputs, as defined in a control algorithm. The DDC controller and its control panel replaces:

– analog controllers

— relays
— amplifiers
— feedback transducers
— setpoint adjusters
— minimum positions switches
— value and status displays
— time clock
— black box EMS controllers
— host-based energy management programs
— control switches
— gauges and thermometers

The relationship of these functional elements is defined in the DDC program.

Since the "digital" computer is basically a binary or "two-state" device, a DDC controller typically includes some integral devices that are external to the "computer" which allow it to do the full HVAC analog control function, i.e., A/D (analog to digital) converters and D/A transducers.

3. *Proportional Control*—A control algorithm or method where the final control element moves to a position proportional to the deviation of the value of the controlled variable from the set point.

For one to understand the higher order analog control modes frequently associated with DDC, a full understanding of this basic (proportional) mode is essential.

Proportional is a control mode where the position of the controlled device (valve, damper, inlet vane) varies proportionally to the value of the process variable (temperature). Most pneumatic controllers are proportional.

A proportional pneumatic chiller thermostat set to produce 44°F water in mid-summer (full load, inlet vanes full open) will produce 42°F or 43°F water at light load. A drop in temperature proportionally closes the inlet vanes.

A proportional mixed air thermostat set to provide 55°F mixed air at 55°F outside (full open outside air damper) will provide 48° to 50°F mixed air when it is 10°F outside. A drop in mixed air temperature is required to proportionally modulate the outdoor air/return air mixing dampers.

Proportional algorithms have two relevant parameters: the "setpoint," and the "throttling range." The proportional range of a thermostat is commonly referred to as "throttling range." A room thermostat set to control a dual duct mixing box from 70°F (full heating) to 77°F (full cooling) would have a 7 degree throttling range.

During system start-up, throttling ranges should be adjusted to the minimum (chiller, mixed air) value at which they will stably control. When they are set too narrow, hunting or two-positioning occurs, often producing unsafe conditions. Wide throttling ranges produce smooth proportional control, but in some cases do not provide satisfactory results.

Conscientious proportional control installers often find themselves readjusting throttling ranges a little wider during start-up and the warranty period due to instability created by load changes.

The proportional effect produces a deviation between the "setpoint" and the actual (control point) temperature. This difference is referred to as "offset." Attempts to reduce or eliminate "offset" sometimes leads us to the PI control mode. The solution to the "offset" issue has been posed as proportional plus integral control.

4. *Proportional plus Integral Control (PI)*—a combination of proportional and a response which continually resets the control point toward the setpoint to eliminate the offset. Also called proportional-plus-reset control and two-mode control.

In this PI control mode, the throttling range parameter of the proportional element of the PI algorithm is usually set comfortably wide to assure smooth proportional action. The "integral" element of the PI algorithm periodically calculates the "offset" value and, if it is not zero, effectively sends a "correction" signal to the D/A device which opens or closes the valve or damper to a position more or less than that set by the proportional element of the algorithm.

Effectively, the integral correction signals accumulate to eliminate any offset created by proportional positioning. The integral portion of the algorithm brings another relevant parameter, the integral "time interval" that elapses between integral effect calculations. This time must be set longer than the time it takes the controlled variable analog input to sense and reflect the effect of the last integral correction signal. This time is dependent upon the controlled system dynamics, the analog input sensor location and responsiveness, and the responsiveness of the analog output and actuator system. Added testing is needed to determine proper effects.

For most applications where the systems being controlled do not run continuously, DDC controllers must include features to prevent "integral wind-up," or the accumulation of the offset signals calculated while the controlled system was down (and nonresponsive). Several methods are used to prevent excessive wind-up. One method is to prevent the accumulative effects of the integral calculations from ever exceeding the value of the throttling range. Another method is to dump the integral signal value upon start-up, and during start-up, suppress the integral calculation until the controlled variable value enters the values encompassed by the throttling range.

5. *Proportional plus Integral plus Derivative (PID)*—A combination of PI action and a response that compensates for the rate of change with which the controlled variable deviates from the setpoint.

The "derivative" response is added to the PI algorithm to help prevent serious over-/undershoot as a controlled variable approaches setpoint; and where there is considerable thermal (in the case of temperature control) mass within the controlled medium, producing a significant flywheel effect in the thermodynamic process. The derivative element of the PID algorithm then calculates the rate of change of the offset, predicts the over-/undershoot, and produces a third (derivative) element of a correction signal which is added to (or subtracted from) the value of the correction signal of the proportional and integral portions of the PID algorithm.

In summary, "derivative" action is proportional to the rate of change. It speeds correction if the control point is moving away from setpoint and slows it if moving toward setpoint.

In most HVAC applications, PI control is better (and much simpler to implement) than PID. An exception is air flow control where experience has proven that, because of the requirement for a narrow throttling range and a very short reset time, a delicately derived "rate" setting can be most helpful in tuning and stabilizing the DDC loop.

6. *Supervisory Digital Control (SDC) (as opposed to DDC)*–A control loop in which a digital controller periodically updates the setpoint of a local loop analog controller as a function of a set of measured control variables and a given set of control algorithms.

An alternative to DDC is for the EMCS to function as an SDC unit. This has been used for chilled water, hot deck, cold deck and reheat unit discharge temperature optimization where the EMCS resets analog thermostats which actually do the local loop control function (usually via separate sensors). State-of-the-art DDC systems are now capable of providing the full control function in a cost efficient manner, and as a result this type of system is becoming outdated. These new DDC systems are also being utilized to replace local loop relay logic associated with safety, interlocking and timing functions. Thus a single DDC system will soon be able to provide global equipment control. To expand on this concept, the upcoming trend is to replace zone controls, such as thermostats, with smart DDC components as well. For example, in Variable Air Volume applications a DDC system would control the primary unit for discharge air and static pressure. Individual smart zone controllers would communicate on a network with the DDC system, while discrete control is provided to modulate zone dampers, sequence internal heat stages and start/stop internal fans.

A key word in both DDC and SDC definitions is "periodically." In the heart of both system computers is a processor executing instructions one at a time. The execution interval of the processor may be a few millionths of a second, but to the process, the computer is usually looking the other way. The interval requirement for a periodic loop update varies greatly between the DDC and SDC applications. In HVAC applications, the SDC loop update interval requirement rarely falls less than five or six minutes, whereas the DDC loop update interval requirement could be as small as a few seconds.

Another necessary comparison between DDC and SDC is the

failure impact on the data environment (DE). Upon failure of an SDC computer, optimization ceases, but the local loop controller usually continues to function perfectly at its last commanded setpoint. The negative aspect of this is that the controlled load may not suffer, and often these failures go undetected. As a result, energy savings are not achieved. This is precisely the reason for a movement towards global implementation of DDC technology. One negative issue which should be addressed with regard to the global DDC approach is that loss of a DDC controller can have serious DE effects like the total loss of a thermostat, and immediate correction of a DDC controller problem is usually necessary. Because of the failure impact and loop update timing requirements, DDC systems usually have far fewer loops per controller than SDC systems.

The advent of distributed processing DDC systems, as discussed in Chapter 7, makes it possible to have stand-alone controllers. In this way, a failure or loss of communication with one loop of control does not affect other loops of control, or the rest of the facility. Certainly, quick response to an equipment failure is still essential to ensure the continuity of space conditioning. However, it is preferable to acknowledge a failure and respond to repair the equipment, than to install redundant layers of control which can have a negative effect on EMCS performance.

BENEFITS OF DDC

DDC has several potential benefits over conventional analog and supervisory types of control. The most significant one, and the one that brought DDC into our everyday lives, is lower first costs. As the cost of microprocessors and memory fell, the percentage of jobs where DDC was the cheapest form of control rose to the point where major control suppliers could economically develop the necessary reliable, flexible, and adaptable DDC systems for routine use.

Certainly, most control loops are still nondigital because of cost; but in many cases, digital is the lowest installed cost. These cost advantages fall into two categories:

1. *Lowest first cost*—Certain unique configurations occur where a single DDC unit can replace complex pneumatic controls and

instrumentation, as in the case of a large dual-duct system with economizer.

Another case in this category would be a job with multiple identical fan systems of moderate complexity where a single programming effort could pay off in repetitive use.

The last case would be a building with two or four simple VAV systems per floor where there is some need for centralized control and/or monitoring. Here, a single controller could perform control of multiple simple identical loops for three floors, and provide inherent central setpoint adjustment.

2. *Lowest life-cycle cost*—In some cases, DDC can be justified as the preferred financial investment by considering its better energy conservation performance and lower maintenance cost, although a pneumatic control system would have lower first cost.

Another benefit of DDC is that it provides a certain degree of data preprocessing which could, in some cases, make a centralized management system affordable, i.e., the DDC controllers have made the analog-to-digital (A/D) conversion and the D/A conversion, and certain sensors are already installed. Although this sounds simple, if the DDC controllers were not designed for interfacing into a central management system, after-the-fact interfacing would be practically impossible. See Chapter 11 on communications and standardization.

Precise central (management) set-point control is the DDC feature that best sets it apart from conventional analog control. The setpoint itself is a binary number within the computer and may be protected from tampering by the use of a software guard. The physical analog temperature sensor is usually a wire-wound or similar resistance bulb whose characteristics are very stable and not sensitive to moisture variations, tampering, or vibration. The most probable sources of error in the DDC system are the sensors' calibration and the A/D converter. A good DDC system will practically eliminate the dreaded "drift" (a mechanical phenomenon resulting in a controller calibration shift caused by aging, relaxing, fatiguing, and deforming of the springs, levers, and bimetals of the mechanically constructed pneumatic type control system), but all DDC systems are not necessarily good or better.

A good DDC system will require less maintenance, and may be further enhanced with internal diagnostics that can report, and in some cases diagnose, component failures. The main issues which act to reduce maintenance are the replacement of the many interconnected analog components with a single programmed computer, and the reduced requirement for calibration checks. This benefit is somewhat offset by the requirement for more sophisticated test equipment, and the need for a systems analyst (one familiar with the total building HVAC process, the control and optimizing strategies, the DDC hardware and test equipment, and the structure of the software logic).

A final benefit of DDC lies in its ability to allow intercommunication between DDC controllers and provide system-wide hierarchical control. An example is for the fan controllers to advise the chiller controller of their demands, so the chiller controller knows what optimum temperature to maintain. Controller-to-controller intercommunication must be designed into the DDC system, and cannot practically be added later without excessive cost.

ADAPTIVE DIRECT DIGITAL CONTROL

Determining the proper settings for control systems has always been a problem and the higher the level of control, the more difficult the determination. With straight proportional control, we just have throttling range to worry about, but an uncorrected, inaccurate setting can cause problems. When integral action is added, we must also determine the reset time which is more difficult than throttling range and also causes problems if the wrong time is selected. When the derivative function is added, the proper rate time setting for the authority of the derivative action is dependent on the characteristics of the system being controlled. To determine these characteristics, it is necessary to run extensive system tests under varying load conditions.

In practice, the settings for the control functions are nearly always determined for one particular load condition; usually full load or design load. If they were properly selected, they will work well for that condition. HVAC systems operate at load levels from full load down to no load, however. The settings that are proper for full load are not necessarily proper for mid- to light-load conditions. The result is that even if all the system testing was done properly and the settings were made perfectly, you end up with a control system functioning at an optimum level "most of the time." If there were shortcuts taken and/or errors made, "most of the time" can degrade to "some of the time." After-the-fact correction of these problems can be a tedious, time-consuming and expensive process. A new technique called "Adaptive Control" can solve these problems, however.

Adaptive control is the latest innovation in sophisticated commercial control systems. With adaptive control, the control system learns from experience and adapts its control "parameters" accordingly. This is particularly important in systems where the process is variable, as in commercial HVAC systems. The digital control "parameters" are the functions which would be "settings" on mechanical control systems. They include setpoint, throttling range, authority, reset time, rate time, etc. There are more "parameters" in digital control systems than "settings" in mechanical control because it is easy to add extra functions in software while each one adds significant cost in a hardware-based system.

Adaptive control is important in commercial control systems because parameters must change as the load on the HVAC system changes. Some characteristics (e.g., coil pickup and sensor response time) vary with the load and therefore, for optimum control all the time, the parameters of the control system should change as well. This is what adaptive control does.

For example: In a variable air volume system, the performance of the coil and the response time of the sensors in the airstream vary with the velocity of the air in the duct. Under full load condition, the velocity is maximum and the temperature rise or fall in the coil will be minimum. The sensors will respond to temperature or other changes rapidly. However, at light load the air velocity will be much lower and the sensors will not respond as rapidly. The temperature rise or fall through the coil will be maximum, assuming the same coil temperature as at full load. Obviously the parameters cannot be the same for light load as full load. Adaptive control continuously senses changes and adjusts parameters accordingly. (Incidentally, adaptive control does not require additional sensors or controls.)

Basically, the adaptive control algorithm determines the ideal process "model" based on current conditions and then adjusts the system parameters (or "settings") as required to match that ideal. The process model is determined by the system response to previous load changes and as that information ages and newer information is accumulated, the older data is given less weight and is eventually dropped out of the calculation. As a result, the model is based on the average of the result of recent load changes with the latest change having the most weight and the oldest the least weight.

Thus when a control system has been controlling in a heavy load situation, i.e., close to design conditions, and the setpoint is changed significantly by a night setback program, the control parameters will have to change because several control system characteristics have changed. (E.g., the coil time constants have changed, and, if it is a VAV system, the sensor time constants have also changed.) The adaptive control system will measure the new system time constants and incorporate them into a new process model. It sends the resulting new parameters to the control equation which calculates the correction signal required to get the control point back to setpoint. The correction signal is sent to the output device, usually a valve or damper actuator, and the process starts all over.

Another type of adaptive control can be applied to energy management programs. Optimum Start Programs function to start the HVAC systems at the latest possible time so they will just reach the comfort range at occupancy time. These programs use factors based on the building design, solar loading, and HVAC system characteristics. They are difficult to determine and are usually established through trial and error, mostly error.

Any modifications or additions to the building or systems will usually change the factors. The adaptive function can automatically determine the correct factors and change them if any characteristics of the building or systems change. It does this by retaining the time the building reached comfort conditions vs. occupancy time from the previous day, and averaging that factor and the corrected factors from previous days to calculate the current day's startup time based on current indoor and outdoor conditions.

Optimum Stop Programs can similarly be improved by adaptive control. This program functions to shut down the HVAC systems as early as possible and still keep the building in control until the end of occupancy. Though the factors may be different, the same problems can occur here as in Optimum Start. Here the adaptive program checks the drift rate after the building is shut down, stores it, and uses it to calculate shutdown time for the next day.

To summarize, the benefits of adaptive control include:

- Eliminates guesswork in setting up higher level control functions.
- Adapts to varying conditions of load, air flow, water and air temperature, etc.
- Provides control equal to or better than PID control without expensive, time-consuming system testing.
- Maximizes energy savings—more so than any system without the adaptive feature.
- Assures that comfort conditions will be maintained during occupied hours—even when conditions change.
- Eliminates the need to revise programming when changes are made to the systems or building.

CONCLUSIONS

The demand, technologically and performance-wise, for DDC are bringing many varied products to the marketplace and will continue to do so in the future. As vendors introduce innovative variations of this continuing dynamic digital technology, it becomes even more vital for the "mechanical practiticners" among us to continually upgrade our expertise. Reliable "control" should be the primary objective of the DDC system, and should never be compromised in the selection of DDC system suppliers.

11

EMCS Communications and Standardization

Over the years professionals in the energy and controls industries have done a great deal of soul searching about communications with automation or Energy Management and Control Systems (EMCS). Out of this process came an industry-wide demand, which may be summed up in one word: STANDARDIZATION. To date EMCS users have been required to support a front end for each manufacturer's EMCS that they purchased. This meant that a user had to learn the necessary language and methods for interface with that particular system, as well as physically maintain the hardware.

This problem was complicated by the fact that these various EMCS would not share data gathered through trend logging, etc. The result of this situation was that users could not carry out their functions efficiently due to a necessary duplication of effort. Also they tended to get locked in with one system, and were less likely to employ new technology due to the burden of adding a new front end, and learning the necessary interface. Standardizing the EMCS communications process is the obvious answer to these problems. However, it is proving to be an extremely complex task. The key schools of thought with regard to EMCS standardization are two: System Level and Communications Level.

The *system level* contention is that all EMCS should utilize a common protocol in transmitting and receiving internal system data for control decisions, data conversion and input/output processing. Ultimately this would allow one manufacturer's CPU to control through another manufacturer's field panel. The drawbacks to this level of specialization are many and include such issues as proprietary manufacturer design, processing speed, and the opportunity for manufacturers to compete in a technology intensive industry.

The *communications level* addresses the fundamental needs which initially fueled the drive for standardization. This communications level option for standardization is much more achievable, as standards already exist for communications. Under this scenario, universal front ends would be developed to allow one piece of hardware and software to communicate with any manufacturer's equipment. This facet of standardization is the key focus of the balance of this chapter which will define in greater detail the process and product of EMCS communication.

This chapter will discuss the management problems which contributed to the need for standardized communications, and the process of EMCS interface. Finally, we will explore solutions which have been posed. It is critical to first establish that this discussion relates to remote communications with EMC systems. Though a distinction exists between all EMCS and those employing state-of-the-art DDC control technology, the basic issue of communications relates to all such control systems. In this case, communications will be defined as system interface, carried out via modem, from a location geographically remote to the controlled facility.

Communication is necessary for EMCS program management. On the most basic level, EMC systems require management to produce the expected energy cost savings. With the current low cost of energy this is even more crucial. Yet in actuality industry records show that these systems are producing benefits beyond the energy savings that were originally conceived. To achieve these benefits in cost avoidance, and to make expanded use of their system, users have developed a new discipline, EMCS Program Management. Among the most effective tools for completing this work is DDC Communication, yet there are a number of complexities to carrying this type of program. These complexities or problems are explained in Chapter 17 on EMCS program management.

Essentially effective program management of building automation programs requires the mobilization of personal computers to serve as "front ends." A front end consists of the necessary hardware and software to monitor and program an EMC system from a local or remote location. In addition to the PC, it will typically include a modem and in some cases proprietary manufacturer interfaces for system interaction.

DDC INTERFACE PROCESS

In Chapter 17 on EMCS program management, a number of tasks and functions are outlined which are necessary for optimal system performance. To explore how these items are completed, this section discusses the process of conventional and optimum interface in more detail.

CONVENTIONAL DDC INTERFACE

To set the stage for discussion of the EMCS communications process, the topic of conventional user interface should be reviewed first. Though the control sophistication of the DDC system surpasses conventional control systems, the communications process is essentially unchanged. Variations on the user interface provided with computer control systems through the years have been designed to mirror the simple capability of a dumb terminal. More complex systems expand on the dumb terminal, and add features to improve the productivity of the process.

Interface systems vary in many respects, but the issues may be viewed through the three basic functions of Front End Interface: System Programming, System Interrogation, and Logging.

The definition of programming varies greatly from one manufacturer to another, but styles may be divided into two categories: Library Algorithms and Free Form Programming. A detailed discussion of EMCS programming is not warranted in this section, yet a key point must be made. The manufacturer's definition of programming must meet with the user's needs. Many users want to be able to generate a complete system program and specific control strategies from a remote location. Many EMC systems will not allow this level of activity via modem communications. If these issues are important to a specific application, ensure that they are covered in the EMCS specification.

Conventional system programming and interrogation functions

include some type of communications features. Communications are typically handled by Site Terminal or Remote Communication.

Site terminals often consist of hardwire interface via an RS232 port, or local dial-up via a modem. The device utilized was often a terminal with limited intelligence capability, and is rarely used for other functions such as office automation, etc.

Remote communications, if provided, are via proprietary front end terminals. These are often specially-configured personal computers or terminals manufactured solely for front-end purposes. In essence communications are provided, though limited, and the user is able to program. Users are provided with access to system instructions, and have the capability to conduct a portion of the tasks outlined under program management.

Difficulties arise because each EMC system has a proprietary set of program instructions. Much of the thrust towards standardization is aimed at developing a standardized protocol to simplify interface with various types and models of EMC systems.

System Interrogation pertains to daily polling operations to monitor facility operations. Among the tasks which may be conducted are: tracking space conditions and point status, diagnosis of problems with systems or controlled equipment, and review of logs. These functions are read only, and provisions are made in conventional front ends for such tasks to be carried out by various members of the user's staff.

Logging features are typically provided so that historical data may be developed on various control and monitoring points. The data may be arrayed by point address, and logged based upon user-defined time intervals.

This issue is crucial to the energy manager, because access to this data is vital to program management. The deficiency in most conventional front ends is that data are provided in specific formats, i.e.: hard copy to printer, and users rehandle that data for reporting.

This involves time-intensive manual sorting of voluminous printouts, rekeying of data into applications programs, or hand analysis and ultimately reporting. The bottom line here is that the time frame for completing these tasks is so prohibitive that managers either never have time for it, or end up making decisions based upon old data.

OPTIMUM DDC INTERFACE

The optimum interface extends the technology we have outlined thus far, and begins to pose solutions to the problems of EMCS program management. This technology builds on the conventional front end by addressing needs and features which are not met by those systems. Communication is the most crucial function available to an energy manager, but the scope of an optimum system must be defined. Provisions must be made for the requirements mentioned in the discussion of program management, and this must be available through local and remote communications.

This discussion will primarily be oriented to the user interface via remote communications. The important elements in this process are a standard microcomputer, and software. It is imperative for users to have the ability to monitor, interrogate and program their systems from a remote location, a single building within a complex, etc.

This issue becomes vastly complicated when users are supporting more than one manufacturer's equipment. At the simplest level the functions mentioned under conventional systems must be provided, however there are numerous other communications features which can substantially improve user productivity. Among these are functions such as: unattended terminal features, dumb terminal capability, communicator files, program mass update functions, and others.

The conventional communications functions are designed for utilization in an attended mode, where the user is a direct participant in the process. They may be conducted in dumb terminal or via a software-defined set of communications instructions and menu prompts. Optimum features, on the other hand, rely on more complex communications capabilities. The steps in this process are based on the need for both attended and unattended communications. Among the essential features are: communicator file capability, communication sequencing or auto-call, and an executive or execution program.

It is first necessary to set up a communicator file consisting of a program instruction set written in the language of the equipment to be interfaced. Note that establishment of a universal protocol would simplify this process for a new EMC system, but an existing system will still require the communicator feature.

This communicator would contain the sign-on, user password,

each of the commands necessary to access desired data, and the sign-off. Communicator sequencing is the process of identifying the systems to be accessed, so that the user can poll groups of systems in succession.

The final step of this process is an executive function which allows the user to define the time for this process to begin. A user can target a group of systems for polling in the afternoon, have the system interrogation completed that evening, and review the data first thing in the morning. The utility of this feature goes beyond data access, however, because it may also be used to define a "task pathway" for any purpose.

Mass update of programs is one example of its utility. Let us assume for a moment that you are using a system which does not automatically compensate for daylight savings time. Simply set up a communicator to make the appropriate change in spring or fall, and have the executive program contact your systems to change the system clock on the designated Saturday evening. Given a little time, energy managers are astute enough to devise unlimited applications for this function.

Finally, it is important to note that this is not limited to EMCS communications. The determining factor must be compatibility with such standards as the VT52 and VT100 protocols. As mentioned under program management, program update is a vital process. The communicator files can also be used to download and upload files to other front ends for update purposes. The same process may be used for in-house data transfer from a regional to a corporate facility, etc.

INTERFACE DIRECTIONS FOR THE FUTURE

In this chapter EMCS communications have been explored and discussed in the real world for real managers. Many of the complexities which have developed through the daily management of EMC systems direct us towards the need for a concerted solution to our dilemma.

In discussing this topic a few years back, the author coined a term "Missing Link" to illustrate the need for a communications link between various manufacturer's DDC systems. Industry-wide acknowledgment of this need has led to the outline of a specific agenda of EMCS interface issues. Link technology is, to a great degree, a response to these issues, and the need for the technology is quite obvious. The need for standardization may be even more clear than before, however the myriad of solutions posed by the problem serve to confuse the issue further.

Key issues regarding the concept of standardization have been defined in this chapter under limited conditions. This is because it is in fact a concept rather than a reality at this point. Rather than leave the concept entirely vague, this chapter will close by exploring options available for resolving the standardization issue. The options available may be viewed in three schools of thought: EMCS network, EMCS software and universal protocol.

In the area of EMCS network communications, manufacturers would develop a local area network format to function in much the same fashion as a minicomputer-based control system. This would allow for global control, security and office automation via one LAN type system. There are some minicomputer-based systems on the market which could fit this definition with limited expansion. This would be an excellent self-contained solution to our communications problem.

However, there are several problems with this format. Networks require long lead times for development, beta testing and implementation. Also, the retrofit segment of the EMCS industry can not make use of this technology without significant capital investment. An additional concern is that users must purchase the LAN, and replace noncompatible controls and systems. Next the user must determine whether LANs will be compatible with one another, or the initial problem is simply recreated on a larger scale. In spite of these issues, it is safe to postulate that the network format is more than adequate to meet the industry need.

An alternative to networking is to accomplish these functions via software designed to run on standard PCs. For the short term this could be effective as an interim or transitional option for several reasons. The initial investment in software should be substantially less. Research and development along with beta testing and implementation are simplified and occur more rapidly. Updates and enhancements to the software are more easily acquired and put into place. Lastly, and possibly most important, the retrofit environment may make immediate use of software technology.

The final, and likely the best possible option, will be developed through steps that are being taken right now in our industry. To provide a well-formulated guideline in this area, the energy and controls industries, through the American Society of Heating, Refrigeration and Air Conditioning Engineers (ASHRAE) have determined that a standard is necessary. In this way, all manufacturers would subscribe to a standardized protocol, and the concerns of program management could focus on optimizing performance, instead of communication.

In summary, users should orient their activities toward the future of EMCS interface, while addressing the needs of the present. In light of the issues outlined in this chapter, it is of optimum importance for users to reevaluate their strategic plans. The program objectives should be revised to account for an EMCS interface standard.

It is important to reassess program management functions, examine the efficiency of current methods, and have input to the link standardization process. Users should define as prerequisites for system interface, that link requirements be fulfilled. In essence, it is of primary importance to establish whether systems currently in use are compatible with the criteria outlined in this chapter.

Determine a plan of action for systems which are not compatible with link technology, based upon the critical nature of control applications and the cost effectiveness of system replacement.

EMCS specifications should be rewritten to reflect the need for universal interface, via established and standardized protocols.

Users should seek out manufacturers who espouse the use of standardized interface, evaluate their product offerings in terms of user requirements, and secure copies of software for use with current systems. Evaluate each component of your in-house program to reflect the state of the art available, to ensure optimum performance.

The product of the efforts outlined above will be enhanced EMCS performance, and simplified program management. The work of the ASHRAE standards committee should be followed carefully, as it will certainly yield a solution for future EMC systems. With cooperation and teamwork, the industry, through manufacturers and users, can develop a standard to unify the EMCS communications process.

12

Designing a Level III EMC System

BASIC SELECTION CRITERIA

The basic criteria to consider in selecting a system are as follows.

Cost Effectiveness

Cost effectiveness is the basic criterion to be used in determining which level of EMC is best and specifically what type of system or architecture will yield maximum return on investment. Cost effectiveness is determined through use of life-cycle costing techniques which are discussed separately in Chapter 19. It is important to note that first cost is not the exclusive criterion. Cost must be weighed in conjunction with the other criteria listed here, to ensure that essential features are provided with the system which is chosen.

Adaptability

How well a given system can be adapted to an existing building is extremely important. Physical requirements of the new system are the most obvious of the adaptability concerns. In fact, will the system fit into the existing space? It is suggested that some 900 square feet are needed for the command center of a Level III EMC system, and that location of the center should be considered strategically in light of the functions to be performed there. If the space is not readily available, can it be made available? If something must be displaced, will new construction be necessary? If so, of course, it must be considered in the economic analysis.

One must also consider existing local controls, or the possibility of installing local controls at this time, with the likelihood that a Level III EMCS will be added at a later date, for example, following completion of planned facility expansion program. In fact, can the EMC system be interfaced with these local controls, or will the controls have to be replaced?

Consider, too, the existing equipment. If much of the most costly mechanical equipment is scheduled for replacement, will the new equipment be compatible with the EMC system? Likewise, if the new equipment will be far more efficient than that which it replaces, is the EMC system needed, or could local controls suffice?

Maintainability

The ability to maintain a system is extremely important. One of the related concerns, discussed below, involves the provider's ability to support in-house maintenance through training programs and manuals. But equally important, if not even more so, is the availability of professionally trained maintenance persons employed, licensed or authorized by the manufacturer.

Wherever possible, references should be checked to determine how satisfied other users of the proposed system are with the maintenance services provided. Key concerns in this regard are completeness of preventive maintenance, responsiveness and capability of outside maintenance crews, the annual cost of the maintenance (which usually is in the range of 8–10% of initial installed cost), and the availability of a service maintenance agreement.

Recognize that, without proper maintenance, the system will not perform reliably. In fact, this has been a critical problem with many Level III EMC system installations.

Ideally, there should be an in-house maintenance capability to provide maximum responsiveness at minimal expense.

In any event, maintainability of the system should be considered in design. While a high level of sophistication may be desirable, it should be recognized that increased sophistication often results in increased maintenance problems and costs. As such, trade-offs may be necessary.

Reliability

Reliability relates to two issues. The first issue is how well the system performs. This information can be obtained primarily from other users of a similar system. If it is subject to frequent breakdown, it has a low level of reliability.

The second issue is the way in which the system performs. If it has been designed and installed well, it should function well so those who rely upon it have confidence in it. On the other hand, the system may be designed in such a way that certain problems are built in so that, for example, occasional interference with transmission lines results in signals that are not received properly or at all, erroneous data, etc.

Reliability can be evaluated much in the manner used for determining the average effectiveness level (AEL) discussed under acceptance testing in Chapter 10. In general, major factors relating to reliability include:

1. The average length of time for which the system performs completely as it should without breakdown.

2. The average extent of downtime.

3. The nature of problems experienced and the degree to which they inhibit the facility from meeting its mission.

Expandability

An existing system may not be suitable to handle additional points. Expansion may require the purchase of another computer, an additional memory and/or more and larger FIDs. It may cause programming inconvenience, high factory charges, and disruption of the normal operational routines. These problems can be minimized by looking ahead at the time of initial design. If a long-range facility plan exists, both existing and proposed buildings can be identified and points allocated at the outset. If there is no long-range plan, an allowance for future growth should be made.

Provisions for future growth affect several elements of the system. The most critical item is the CPU, its memory, and its software programs. Equipment purchased either should have spare capacity for all future growth or should be modular, permitting easy addition of more points. Points cost approximately $500 to $700 each at the time of initial installation; $1500–$2000 when added later.

Field wiring and the FIDs also can be designed to minimize the impact of future expansion. If a coaxial cable or twisted pair wiring is used between FIDs, no additional wiring should be needed, assuming all FIDs were purchased initially. Problems may arise if the FID has not been specified and/or furnished with adequate spare capacity. It is suggested that FIDs be selected to handle all known future points, plus 20% expansion capacity.

For known future buildings, air handling units and related equipment can have space reserved on the CRT format and then have information field programmed when the anticipated points are added.

In some cases, suppliers will provide unit prices on items for future expansion at the time of the original bid. With an agreed-upon cost index or escalation factor, all future additions should be easily priced using the unit prices given.

Note that the impact of time must be considered in all these activities. Since the life expectancy of a Level III EMC system generally is considered ten years, due to the rate of state-of-the-art advancement, it would not be practical to specify spare capacity for expansion likely to occur more than ten years after initial installation.

Programmability

It is recommended strongly that the system be user-programmable. In this way, programs can be easily modified or added in the field as needed. User-programmability is recommended to eliminate reliance on manufacturers for expensive program revisions. The EMC system consultant should perform or be consulted extensively for programming. His involvement in this procedure will help ensure that programming requirements and system design are compatible.

TECHNOLOGY ISSUES

These issues may be viewed in three categories: Technology Requirements, State-of-the-Art Technology, and Communications.

Under *Technology Requirements,* it is essential to determine that the specific needs for each particular system are evaluated. Sophisticated or unused features which are provided with an EMC system, but are not implemented, result in unnecessary cost. Conversely, acquiring a system which cannot do optimized stop/start for an application where this is a critical feature would be a mistake. Securing a qualified automation consultant and careful evaluation are necessary to ensure that technological requirements are met.

State-of-the-Art Technology is related to the requirement issue. In this case, the user evaluates the quality of the features provided. A number of industry buzzwords are used in promoting systems, and each of these is discussed in this book. State-of-the-art becomes an issue when an application requires the sophistication of DDC technology or the stand-alone benefits of distributed processing. There are numerous EMCS designs and the focus here is on matching appropriate technology to a given control situation.

The final technology issue is *Communications.* An entire chapter has been devoted to it, therefore only cursory treatment is necessary here. The primary concern is that communication needs be defined for each application. Once this process is complete it will be possible to determine whether a given EMCS can meet those needs. Refer to Chapter 11 for an overview of the points that should be addressed under this heading.

DRAWINGS

Once a system has been decided upon, drawings and specifications —the contract documents—must be prepared.

Drawings are extremely important. They must be reviewed carefully for completeness and accuracy, and should clearly and precisely indicate the designer's intent. It is also very important that the consultant carefully coordinate the drawings with the specifications in terms of the component's function with respect to each other.

Many things should be included in a set of drawings. These are indicated in Table 12-1. A few of the most important factors, basic to all EMC system packages, are:

1. Block diagram,
2. interconnecting trunk wiring data link diagram,

Table 12-1. Drawings Needed for Level III EMC System

1. Title Sheet
2. Drawing List
3. Legend and Symbol List
4. Site Drawings—Buildings and Data Transmission Cable(s) Routing
5. Data Transmission System Network Diagram
6. Interconnecting Trunk Wiring Data Link Diagram
7. Point Schedule
8. Functional Layout
9. System Schematic Diagram
10. Typical Control Wiring Diagrams—Interfaces and Interlocks
11. Typical Control Sequence Description
12. Floor Plans—Each Building or System
    a. Sources of Available Electrical Power
    b. Location and Nameplate Data of Equipment to be Monitored and Controlled
    c. Location and Type of Existing Controls and Starters
    d. Location of FIDs and Data Transmissions or System Terminations
    e. Location of Radio Equipment
    f. Related Mechanical/Electrical System Modification
13. Layout Drawings
    a. Typical Aerial Data Transmission Wiring
    b. Typical Underground Data Transmission Wiring Details
    c. Data Transmission Wiring Building Entrance Methods
    d. Detailed Power Wiring from Available Electrical Sources
    e. Typical Radio Equipment Installation Details
    f. Typical FID and MUX Installation Diagrams
    g. Typical Instrumentation Installation Details

3. point schedule,
4. functional layout diagram,
5. system schematics diagram,
6. proposed locations of FIDs and MUXs,
7. equipment modifications drawing, and
8. existing control system and interface diagram.

Block Diagram

The block diagram of the entire system identifies all major components and how they relate to each other (System Architecture).

The diagram should show all remote-control installations, communications links, interfaces, and all central equipment.

Interconnecting Trunk Wiring Data Link Diagram

An interconnecting trunk wiring data link diagram (Figure 12-1) is required when more than one building is involved. The system diagram indicates the type of communications links used (coaxial cables, microwave links, telephone lines, etc.) and how they relate to each other.

Point Schedule

The point schedule or point matrix (Table 12-2) for the entire system describes each piece of equipment in a finite manner: What it does, whether it is on or off, whether it is a reset, whether the system is monitoring, alarming, printing, etc. The point schedule ties together all the words in the specification, the diagrams on the drawings and the actual functions.

It is generally preferable to place the point schedule in the drawings rather than specifications. Drawings usually are saved; specifications are not.

Functional Layout Diagram

The functional layout diagram relates to the area where central components are located. The layout should consider how the system will be used and who will use it or have access to it.

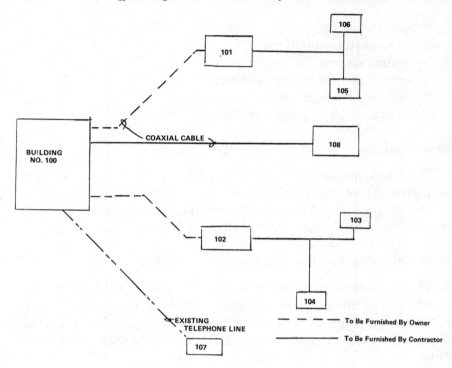

Figure 12-1. Interconnecting Trunk Wiring Diagram

System Schematics Diagram

A system schematics diagram (Figure 12-2) details schematically how specific HVAC systems are to be monitored and controlled.

FID and MUX Locations Diagram

This diagram shows the number and locations of FIDs and MUXs. Note that FID and MUX locations may not be fixed at the time the drawings are made. In such a case, alternative locations should be indicated on the drawing to permit the contractor to make a more accurate cost estimate.

Equipment Modifications Drawings

Equipment modifications drawings are provided to indicate

Table 12-2. Point Schedules

Category	Subcategory	Point Description													
PROGRAMS		Optimized Start/Stop Enthalpy Optimization Duty Cycling													
ANALOG POINTS	VALUE INDIC.	Temp. °F, DB Outside Air Motor Amps													
		Pressure, PSIG KW Demand (High Alarm Only)													
	VALUE INDICATION WITH HIGH-LOW ALARM	Water Temp., °F: Chilled Supply Condenser Supply													
		Rel. Humidity, %: Room RA/Exhaust													
		Air Temp. °F: Room RA/Exhaust Supply Air Cold Deck													
		Totalized Run Time: Hours Calendar-Date Time: Days													
		High Pressure Drop: Prefilter High Press. Drop: Absolute Filt.													
DISCRETE POINTS	STATUS ALARM	High Temperature Low Temperature High Pressure Low Pressure Low Vacuum Low Liquid Level Battery Charger Malfunction													
	STATUS IND. W/ALARM	On/Off Fast/Slow/Off Open/Closed Flow (Diff. Press.)/No Flow On (Supplying Power to Load)/Off 100% OA Mode/RA Mode (Local SW)													
DISCRETE AND ANALOG POINTS	CONTROL	Start/Stop Auto Control/Off Start Fast/Slow/Stop Reset Cold Deck Temp. Reset Hot Deck Temp. Reset Chilled Water Sup. Temp. Reset Cond. Water Sup. Temp. Open/Close 100% OA/Min. OA Acstatus Green/Red Lights (Local)													
		SYSTEM OR EQUIPMENT													

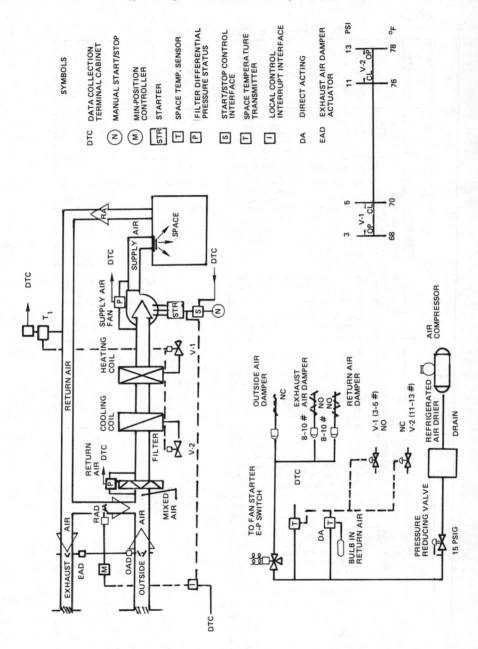

Figure 12-2. HVAC System and Control Schematics

changes which will be made to various pieces of mechanical and electrical equipment to optimize their function. These relate to modifications of ductwork, dampers and damper motors, control valves, contactors, etc.

Existing Control System and Interface Diagrams

Existing control system and interface diagrams are extremely important because they delineate where all the existing components are located and if for example the system is to perform motor start/-stop, the contractor must know where the motor starter is located, what kind of starter it is, and what he must do to that starter to include it as part of the entire system. Everything must be shown explicitly, detailing what can be used, what must be replaced, and what must be added. Showing a box with an "X" in it, or just writing "Enthalpy Control" or "Provide on/off Auto Control" is not sufficient. Providing detail is necessary for bidder information, and to safeguard against a "blank check" situation caused by changed orders needed to correct problems resulting from lack of information.

13

Specifying a Level III EMC System

Presenting a complete guide specification for a Level III EMC system is beyond the scope of this book. Appendix D includes references which can provide detailed specifications.

Two sample specifications which have been used successfully are included in this chapter. They *are for guidance only and are not intended to be used verbatim, for any purpose.*

The system being specified may represent only the first stage of a proposed long-term effort which eventually will result in a comprehensive system in accord with a long-term plan. In such cases, those developing the system should do so in recognition of long-term goals and objectives. Likewise, if an existing EMC system is to be incorporated into design, system designers must do so with existing systems requirements and restrictions in mind.

Development of a specification, especially for a large system, requires consideration of many interrelated factors. Failure to make these considerations may result in a considerable amount of wasted time, effort, money, and energy. For large systems, it is recommended that the specification be developed as a team effort, with team members including, but not necessarily being limited to, mechanical (HAVC) engineer, electrical engineer, electronics engineer, communications specialist, chief operating engineer (or equivalent), contracting officer, and legal staff or counsel. These same persons would evaluate bids and proposals submitted by offerors.

Coordination of a specification with the drawings is mandatory. If materials or equipment are to be furnished under one section, but installed, connected, or placed in operation under another section of the specification and/or the drawings, *state that fact clearly and concisely* in all sections involved. Each discipline should review the entire specification to ensure that language is included to provide complete and operable systems and equipment.

When an addition is to be made to an existing EMC system which uses proprietary software provided by the initial EMC system supplier, a certified letter must be sent to that supplier giving notice of the project addition. The letter must solicit the supplier's participation in bidding on the project addition. Also, the low bidder, if other than the original EMC system supplier, must sign a statement to the effect that disclosure of the proprietary information or use of such data will not be made.

MAJOR SPECIFICATION CONCERNS

Degree of Accuracy

The degree of accuracy expected from a system must be specified. Assigning given degrees of accuracy to components is not enough. The specification must apply to the system as a whole. If the accuracy of a component is within ±5%, and a signal passes through six such components before being projected on the CRT screen, the resulting inaccuracy may turn out to be as high as ±30%. If such a number is then fed into a computer to calculate Btuhs, the result could be unreliable.

Be realistic in estimating the accuracy possible in the particular system being designed. For example, a signal may travel through a mechanical component—a valve, a damper operator, or a meter—then be changed to an electronic signal and travel through a processor or over a communication line and then back to the central component. In that case, it may not be reasonable to expect the accuracy to remain within a small percentage.

Level of Access

Level of access to the system also should be specified. To be sure that a system will be used properly, access must relate to operators' level of expertise. Accordingly, the specifier must decide who should have access to the program in order to change some items, or whether programmers should be hired to service the system. The specifier also must ascertain what checks and balances are required so an untrained operator cannot inadvertently rewrite part of the program or wipe out part of core memory without being aware of what he is doing. The specifier therefore must specify levels of access as well as the software and hardware to accommodate those levels.

Use of Existing Controls

Disregarding the discussion of stand-alone distributed processing systems, there are many systems both installed and on the market today that utilize existing facility controls. In determining the type of system architecture which is to be installed, specifications must cover the use of existing controls. In general, it should be the responsibility of the supplier to inspect existing controls to determine which can and cannot be used, based on the system the supplier intends to install. The evaluation should consider not only the adaptability of a given existing control, but also its condition. If the control can be used only if modified, it generally is best to replace the control, since modification may cost almost as much as replacement. In this regard, items to consider include the cost of replacement, the cost of modification, and the usable remaining life of the control subsequent to modification.

If an existing control is used, but fails during acceptance testing, the cost of repair or replacement should be borne by the facility, unless it can be shown that more adequate inspection by the contractor would have resulted in control replacement to begin with. In that event, the contractor should be liable for the replacement costs, less the cost of the equipment and work that would have been required to install a new control device to begin with.

If a modified control fails, the contractor should assume all responsibility because he indicated that modification was desirable, and the contractor was responsible for making proper modifications.

Electric Power Consideration

Specifications must cover details of the electric power supply and the extent of variation in electrical power which the contractor must provide for. The sensitivity of central processing and computer equipment are such that uninterruptible power supplies or some line conditioning on the main power may be required to guard against voltage spikes.

Factory Debugging

Specifications should require the manufacturer to test all appropriate equipment and programs before they are delivered to the site. The specification also should require the manufacturer to provide written records of the test reports for each component of the system.

Acceptance Testing

Specifications should detail acceptance testing procedures.

The final operational test should be conducted on a 30-day, around-the-clock basis, once the entire system is installed. To determine whether or not the system passes the test, an average effectiveness level (AEL) should be established. The AEL should be expressed as a percentage of time that the system is operating completely as it should over the 30-day, 720-hour period. It is suggested that a minimum AEL would be 95%, excluding downtime caused by:

1. A power outage, providing that all related backup functions related to power outage functions properly,

2. failure of a communications link, providing that the field interface device operates automatically in the stand-alone mode and that the failure was not due to provider-furnished equipment,

3. failure of pre-existing equipment, or

4. a functional failure due to an individual sensor or controller, providing that the system recorded the fault and that the AEL of all sensors and controllers is at least 95% during the test period.

A minimum of two persons generally are required for a test: one at the CPU to operate a point and command it to program its various functions; the other to observe the point and to ensure that the function has been carried out properly. *Every point in the entire system should be checked* in this manner.

Testing also is needed to confirm, for example, that the temperature and pressures read are being read in fact, and that they are being read within the tolerances required. Likewise, alarms must be checked to ensure that they are operating properly.

Although it is necessary to check each point, it is not necessary to check each point each day of the test period. However, random testing of each type of point should be conducted each day of the test period.

In the event that the system does not pass the AEL, the testing period should be extended on a day-to-day basis until the required AEL—95%—is achieved.

Most testing should be conducted in the presence of the contractor, the operating engineer, and the EMC consultant. The satisfactory completion and documentation of these tests will be followed by system acceptance.

Standardization

Specifications must deal with the issue of controls standardization, particularly from the perspective of remote communications and data access. In Chapter 11 this issue was discussed in detail. However it is essential to address this concept in the specification context. The primary concern is that specifications outline communication needs for each application.

The sample specification in this chapter describes these issues in general terms. Users should define requirements with regard to their situations, such as:

- hardware compatibility,

- modem types and baud rates,

- EMCS data conversion for use with particular applications programs, etc.

It is also important to verify that these requirements are met prior to final EMCS vendor selection. Once this process is complete it will be possible to determine whether a given EMCS can meet the specification and the user's needs.

SAMPLE SPECIFICATIONS

Note: The information given here is provided for guidance only. Under no circumstances should either specification be used verbatim without customization by an engineer for the specific application. The authors do not assume any liability for the use of this specification in any way. An individual specification must be a team effort, as discussed earlier in this chapter. It is advisable to have any specification you develop reviewed by your legal staff or counsel.

SPECIFICATION 1 – Instructions to Bidders
(Provided by George R. Owens, *P.E., C.E.M.)*

This work is part of a more detailed specification, and does not include any general conditions—insurance requirements, warranties, workmanship, code requirements, etc.

This specification is to obtain costs to purchase and install an Energy Management and Control System (labeled EMCS throughout this specification). The EMCS will be installed for the purpose of controlling energy-consuming devices with the intent to reduce overall consumption.

The EMCS shall be provided to include, but is not limited to, the following: The program data shall be designed to monitor, interpret and act on dynamic data provided by the system's sensors and controllers. It must be designed to accomplish:

- Reduce kilowatt hour consumption through Time Program and Duty Cycle control of HVAC and associated mechanical equipment.

- Reduce electrical demand charges through peak demand forecasting and load shedding.

- Reduce fuel consumption by reducing heating and cooling load on HVAC units through warm up/cool down cycle of OA and RA dampers and enthalpy control of dampers.

- Optimized system startup and shutdown based on weather conditions, internal building mass, HVAC system dynamic response, and space temperature conditions.

- Reduced energy consumption through reheat reduction and/or unit discharge reset.

The systems shall also be capable of monitoring, gathering, calculating and recording data associated with the management of energy. Included in this capability should be degree-day calculation for both heating and cooling days.

The EMCS specified under this section shall be totally solid-state using digital technology to insure long life and low maintenance costs. The system must be a standard with the manufacturer to insure on-going parts availability and trained technical support. The initial installation shall include all pushbuttons, indicators, switches, pilot indicators, digital and analog value displays, transmission line interface equipment and software, etc., to make up a completely operable system. The initial installation shall have the capacity to handle the points specified in the Energy Management Point Charts plus 25% additional. The EMCS shall be designed in a modular fashion to insure future expansion capability whether it be additional data gathering panels (DGPs) or central console function capability. The EMCS is specified herein to help insure proper and efficient utilization of the mechanical and electrical systems.

Equipment and material furnished and installed by contractor. The EMCS system furnished and installed by the automation contractor shall be complete in accordance with all specification requirements and shall include all equipment and materials to accomplish the objective listed above. Any feature or item necessary for complete operation, trouble-shooting, and maintenance of the system in accordance with the requirements of this specification shall be incorporated, even though that feature or item may not be specifically described herein. This shall include hardware and programming documentation such as logic diagrams, data and software flow charts, program listings and operator manuals.

CODE APPROVALS

All wiring shall be in accordance with the appropriate sections of the National Electrical Code. It is the intent of these specifications to define a single integrated system that will have the capability of monitoring and controlling all mechanical systems both present and future. The system must be UL listed to meet code requirements. Also, all national, state, county and city codes must be met for the locations to be installed.

1.0   BIDDER PROCEDURES

Supply costs for purchase and installation of the EMCS described in this specification. All provisions of this specification shall be met unless excepted in writing with the bidder's proposal.

1.10   FIRM PRICE BID

The firm price bid will be based on the following:

1.11   CPU (Central Processing Unit) or Distributed Processing Units with minimum capabilities of handling (   ) locations with (   ) utility meters.

1.12   All Software costs for the following programs
   A. Start-Stop
   B. Fan Cycling
   C. Enthalpy measurement and control
   D. Control point reset
   E. Demand limiting by adjusting the fan cycle routine and load shedding
   F. Management reporting
   G. Alarms
   H. Custom Programming capability

1.13 Operator's terminal

1.14 Portable operator terminal that can modify system operation from the remote locations and the Energy Engineer's office. One wide carriage printer shall also be located in the Energy Engineer's office.

1.15 Data gathering panels at the remote locations that sense contact closures, temperatures, humidities, etc. and converts these signals to a signal for telephone line transmission.

1.16 All remote sensors

1.17 Complete installation of all the equipment, wiring and sensors, as described in point description sheets and drawings.

1.18 Factory checkout of all equipment prior to shipment.

1.19 Complete checkout and in-service proof of operation.

1.20 One year warranty on the complete system including all equipment and the installation. Warranty period shall start after the checkout and proof of operation.

1.21 Training to be provided:
Programmer training, 2 people on site or remote
Operator training, on site, 3 people

2.0 SCHEDULE

Issue specification	(Issue date)
Bids received by	30 days after issue date
Letter of intent	120 days after issue date
Start construction	210 days after issue date
End construction, start checkout	300 days after issue date
End checkout, turn over system	330 days after issue date

The above is for the firm bid portion of this specification.

DETAILED SPECIFICATIONS

SPECIFICATION 2 – Detailed Specification for EMCS
(Provided by George R. Owens, *P.E., C.E.M.,* and Terry Simms)

3.0   GENERAL

This section of the specification provides general information for the contractor giving the location, installation requirements and a brief description of the mechanical systems of the project.

A very important point in the installation section is that this project is to be a "turn-key" installation. The contractor is to turn over to the owner a completely installed, programmed, checked-out and running system.

When writing a specification, the goal is to write specific enough to insure that the system is more than a time clock while still being general enough to insure a competitive bid.

3.0   GENERAL:

A. *Location*–This specification describes a microprocessor-based Building Energy Management Control System (BEMC) to be installed (specifier to insert location, address, contact person, etc.)

B. *Installation*–The BEMC contractor will be responsible for the installation of the base BEMC System and for the connection to all existing equipment described in the point list. The BEMC is to be a "turn-key" installation. Contractor will install a complete, fully functional System including all labor, supervision, components, materials, hardware, software, warranty and training and is to be responsible for initial startup, data entry and System operation and warranty period service.

C. *Physical Data*– Most HVAC equipment including fan coils, AHUs, exhaust fans, and baseboard heaters, is distributed throughout the occupied areas of the facility. (Specifier to identify the location of major mechanical and electrical rooms.)

D. *Connections*—All cooling systems, air handling units, fan coils, and exhaust fans are to be connected to BEMC System by the contractor as described in this specification and in the point list. The electric baseboard heaters will not be connected as part of this project but BEMC System output point capacity is to be available to connect this equipment at a later date. The common area and parking lot lighting systems are also to be connected to the BEMC by the contractor.

3.1 ENERGY MANAGEMENT SYSTEM FUNCTIONS

A brief description is given in this section to tell the contractor the type of system being installed. Note that the complete range of Energy Management System software is required and that the system must have analog input/output capability.

Important points of this section are: 1) off-site communication feature which allows either the installing contractor or a remote owner to monitor the system without going to the project. This has been important in optimizing the performance of Energy Management System's savings. 2) 24 hours of data storage requires a floppy or hard disk storage medium.

3.1 ENERGY MANAGEMENT SYSTEM FUNCTIONS:

The System is to be installed for the purpose of managing, monitoring and/or controlling certain building HVAC and lighting equipment.

The System is to provide fully automatic Time-of-Day Scheduling, Adaptive Optimal Start routines for both heating and cooling systems, Heating Setback, Temperature-based Duty Cycling and Demand Limiting/ Load Shedding control for assigned equipment.

In addition to its primary function as an automatic building HVAC and lighting systems operational manager, the BEMC is also to serve the secondary function of reporting on and providing data about the mechanical system's operating conditions for purposes of optimizing its performance with the goal of reducing energy use and cost.

The BEMC shall be capable of remote off-site communications and monitoring through the installation of an on-site telephone interface (modem) and dedicated, voice-grade telephone line.

The BEMC shall also be capable of automatic storage of input point information and events for at least the previous 24 hours.

3.2   PHYSICAL DESCRIPTION

A.   *Operator Console* – Minimum requirements are a CRT and printer. L.E.D. readouts are not sufficient for a system this size nor are they acceptable.

B.   *Central Processing Unit* – A computer is required that will handle all of the input/output points and the full range of software options. The expansion capability of the system is extremely important. The absolute amount of memory is unimportant; however, the number of additional points and software programs that can be added is extremely important. Don't box yourself in!

3.2   PHYSICAL SYSTEM

The BEMC shall be a microprocessor-based, central supervisory System including no less than the following components:

A.   *Operator Console* – This console shall include at least a monochrome monitor and a dot matrix printer and must be capable of providing no less than a printed hard copy of all data entry and data acquisition including automatic printing of System alarms.

B.   *Central Processing Unit (CPU)* – The central processing unit shall be a state-of-the-art, microprocessor device provided by the BEMC System equipment supplier and shall be capable of running software required to provide all specified facility energy management functions. Sufficient memory (RAM and ROM) shall be provided to store the entire operating System as well as input data from binary and analog sensors required for managing facility energy consumption. A 50% increase in CPU point capacity must be achievable at no additional cost to provide for future expansion as needed.

The CPU shall be uploadable directly from a disk drive and remotely through modem/ voice line link.

The System shall also have sufficient Input/Output ports to support all System peripherals listed as well as a disk drive. The CPU shall have inherent capability for future expansion without removing, replacing, or obsoleting any existing System components.

C. *Remote Interface Devices—* "Smart" remote or field interfaces are described. However, also note that a central system with hard wiring back to the computer is also acceptable.

It is the authors' experience that both systems work equally well and either one may cost less depending upon the actual building characteristics.

C. *Remote Interface Devices (RIDs)—*Remote or field Interface Devices are defined as the remote, distributed, processor components which directly operate the System output devices (relays) and/or receive input signals from the System sensors. These devices, *when included in the System architecture,* (i.e., distributed network) should be capable of independent time-of-day scheduling control and remote setpoint (temperature) control in the event of loss of communication with other RIDs or the CPU.

RIDs which are associated with HVAC equipment operation are to be installed in Nema 1 or Type A electrical enclosures and may be installed in an approved, non-public area as best suits the overall System installation. At least one spare input and output point shall be available at each RID location and the System shall have inherent capability for future expansion to add new points of control by adding additional RIDs. Any such future expansion must be able to be accomplished without the removal or replacement of any existing RID or equipment related to the RID.

D. *Input/Output Devices—*
Provides a description of the relays, sensors and utility meters that are the interface to the real world. Again, the specification assigns the responsibility of providing a complete system to the contractor.

D. *Input/Output Devices (I/O)—*
All required Input/Output (I/O) devices are to be installed at the appropriate point of measurement or control and hardwired to its appropriate RID or the CPU as dictated by the System architecture.

Remote isolation relays may be installed in Relay Panels or enclosures located as near to the controlled device as possible. They may also be installed in existing electrical cabinets and unit control panels but should be clearly marked and should not interfere with access to, or operation of, the native controls except as required.

New input and output devices provided under this specification shall be compatible with the BEMC hardware and firmware, and shall provide accuracy commensurate with the application. They should also have a record of reliability and performance in similar applications.

The BEMC contractor shall be responsible for providing all components, sensors, signal manipulators and converters required to deliver all analog, digital or binary information as called for in the point lists to the Remote Interface Device and/or Central Processing Unit. This includes provisions for continuous monitoring of the facility electrical power service. The BEMC contractor may accomplish this by using the electric utility meter output(s) but all installation and start-up costs imposed by the utility must be included in the quoted price.

E.   *Manual Override Switches*—The authors will not allow a system to be installed without these override switches. It is not a question "if" these switches are required but "when." The operator must be able to override the system upon a failure. Comfort and safety come first.

E.   *Manual Override Switches*—A manual override capability shall be provided for each output load or load group. This override may be located at the CPU, the RID or the controlled device but should be absolute and should completely remove the controlled load(s) from the BEMC System control regardless of point status.

F. *Wire and Cable*—The BEMC contractor will be responsible for *all* wiring between RIDs and I/O of permanently installed equipment and systems, as well as the wiring for the communications network between RIDs and the CPU.

Transfer of data and instructions between the CPU and RIDs (communications) is to be accomplished through dedicated multiplexing or duplexing cable. Contractor is to use plenum cable or approved flame spread protective cable for all low voltage wiring in concealed wiring runs and in all return air plenums.

Communications cable shall be shielded cable or twisted pair, low voltage wire as recommended by System supplier and shall meet all local fire and safety codes.

Low voltage control wiring shall be at least 16 gage stranded copper and shall be protected by conduit when exposed to weather or other destructive forces or when required by code. Thermostat wire or cable is *not acceptable*.

It will be the responsibility of the contractor to determine the exact nature and quantities of interface devices, electric relays, PE and EP relays, signal converters, transducers, sensors and transmitters required to meet the functional and operational performance criteria set forth in this specification.

G. *Telephone Interface–*
Important for monitoring the function of the system from a remote location.

G. *Telephone Interface–*The System shall be capable of remote off-site communications (uploading and downloading) and monitoring through the installation of an on-site telephone line interface (modem) and dedicated, voice-grade telephone line. All data entry, data acquisition, and data logging operations should be accessible via this remote communications link. The communications software for this BEMC shall be included in the System as bid.

3.3 BEMC ARCHITECTURE AND SOFTWARE

The key statements of "menu-driven" and "self-prompting" in this section are very important to a successful E.M.S. The operator is the most important player and if the system is hard to use, the savings will not be fully realized.

3.3 BEMC ARCHITECTURE AND SOFTWARE

The Building Energy Management Control Unit to be incorporated into this System shall be state-of-the-art, distributed network, modular architecture, or Central Processing with dumb, downline field wiring panels, or a combination of the two. In either case, a central processor shall be included to manage all peripheral equipment, Input/-Output, data storage, communications, and special energy management operating instructions through the System software.

The System software will be menu-driven and self-prompting and shall utilize standard English language commands and prompting. Data entry for monitoring and time schedules, alarm limits and operating strategy assignments shall also be performed using standard English and numeric values (not **BASIC** or other programming language or low level compiler or machine language).

3.4 OPERATIONAL STRATEGIES

The real heart of the system. These software applications provide the energy saving opportunities. Without these, the system would be an expensive pile of wires and silicon.

A. *Alarm Monitoring*—Alerts the operator to problems and hopefully solutions before they become critical.

B. *Time of Day*—This time clock function provides the greatest reduction in an otherwise uncontrolled building.

3.4 OPERATIONAL STRATEGIES

The basic operational strategies and functions to be performed by the BEMC are to include at least the following:

A. Alarm (high/low temperature) monitoring of all System binary and analog input points. Alarm points for analog inputs are to be entered.

B. Time-of-day scheduling control (ON/OFF) of all equipment and systems as indicated in Point List. Scheduling strategies shall include multiple daily on/off times for each day of the week as well as special holiday schedules. System shall be capable of providing a discrete weekly schedule for each controlled load group as defined in the point list.

C. *Adaptive Optimal Start*—This program provides the next greatest opportunity for a reduction in energy cost. Starting the HVAC equipment in time for occupant comfort but no sooner results in up to 4 hours reduction per day.

C. Adaptive Optimal Start routines for heating and cooling incorporating both indoor and outdoor temperature into the optimal start-time calculation. The start-up advance should be self-adjusting to actual building thermal characteristics. The System is to be capable of allowing a user-entered equipment operating strategy for the morning warm-up period including positive shut-off of all outside air dampers, exhaust/-return fans, and economizer operation and a recirculation-only operating mode on air handling units to transfer heat to interior spaces.

D. *Duty Cycling*

D. Duty cycling based on temperature for each listed HVAC unit, exhaust fan or ventilation system.

E. *Demand Limit/Load Shed*—
The single most important concept in making these strategies work is *Temperature Compensated*. Unless the system is sensitive to occupancy comfort, the first time a complaint develops, the operator will tend to override the system.

E. Demand limit/load shedding control strategy to reduce building electrical demand during peak demand (kW) situations. kW signal to BEMC for monitoring and demand control routines will be either from the existing power company meters or through independent measuring devices, i.e., current transducers (CTs) and totalizer. Where there are two electrical services serving the facility, both services shall be monitored for kW and kWh.

F. *Analog Light Sensor*—This fairly new innovation allows the EMS to look at the actual light level outdoors and control lighting accordingly. For example, the interior lights near a skylight or atrium might be turned on at 50% light while the parking lot lights would be turned on at 10%.

G. *Setback Control*—By controlling hot or chilled water temperatures, a reduction in system energy usage can be achieved. It takes less energy to produce slightly warmer chilled water and slightly cooler hot water while still satisfying comfort conditions. These points tend to be expensive, therefore only specify them where truly justified.

F. An analog, ambient-light-sensing input is to be provided by the BEMC contractor. This input is to be linked through System software to disable certain lighting circuits under selected adjustable, ambient light level conditions.

G. Setback control to reduce heating for the building during unoccupied periods. The indoor temperature input for this strategy should be user-assignable from any or all of the indoor temperature inputs called for in the Point List. The outdoor temperature should also be incorporated into this operating strategy as an override to disallow heating System operation entirely during unoccupied periods during mild or warming weather conditions.

H. Data Logging and Reporting capability shall be provided for certain data points and events. This includes storage of the peak electrical demand for the period and total accumulated electrical consumption for the period. The System should also be capable of automatic reporting of selected operating conditions such as temperatures and status inputs, a minimum of three preassigned times during each 24 hour period.

I. Building purge cycle through the economizer System to remove excess, accumulated heat prior to mechanical cooling System start-up.

3.5 **EQUIPMENT INPUT/ OUTPUT REQUIREMENT**

Each of these generic descriptions is self-explanatory. However, this section represents over 60% to 70% of the cost of the EMS. Therefore, choose your points wisely. Too many and the cost of the system is prohibitive. Too few and the system will not achieve the savings as anticipated.

3.5 **EQUIPMENT INPUT/ OUTPUT (I/O REQUIREMENTS):**

The Point List contains a complete list of all equipment and systems in this facility to be controlled by the proposed Building Energy Management System (BEMC). The Point List also indicates the System features (programs) which apply to each piece of equipment as well as the quantity, function and type of related input and output devices (sensors, transducers, signal converters, switches and relays) that are required.

A brief conceptual description of the general Input/Output strategy for each type of generic equipment is provided below. These descriptions are for reference only and represent minimal instrumentation and control. Actual input/output requirements for this project shall conform to the Point List which may exceed the requirements below. Also, the Point List may contain equipment not described below and, conversely, some of the generic equipment and System types described below may not be applicable to this particular project.

A. *Packaged Cooling/Heating Units:*

Inputs—For units 10 tons and larger, provide two analog temperature inputs per AHU: one to sense supply air temperature and one to sense return air or space temperature. For units of less than 10 tons, provide one analog space temperature input per control group, as indicated in Point List. Each group should be limited to 20 tons total with no more than 4 units per group.

Outputs—Each unit is to have at least two remote interrupt relay connections to the BEMC. These relays shall be capable of turning off and restarting the unit fan and cooling functions and disabling the heating function independently subject to limitations of the native control System.

B. *AIR HANDLING SYSTEMS*

Variable Air Volume (VAV) Air Handling Units

Inputs—For AHUs over 4,000, CFM provides two analog temperature inputs per AHU; one to sense supply air temperature and one to sense return air or space temperature.

Outputs—Each AHU shall be provided with a master interrupt relay connection to turn off and restart all System functions. In addition, each AHU shall be provided with separate interrupt relays to disable the heating function independently. In the case of AHUs with economizer cooling capability, a separate relay shall be provided to disable this function independently. In all cases, the outside air dampers shall close upon unit shutdown. On systems of 10,000 CFM and up with multi-stage cooling coils, relays shall be provided to disable each stage of cooling independently.

IMPORTANT: Make sure that no safety interlocks or life/safety system interconnections are violated.

Note: Care must be taken to avoid defeating the native-control freeze, smoke and fire protection strategies with the connections to the BEMC. All these functions must remain fully operable regardless of the status of the BEMC relays and outputs.

Exhaust Fans

Inputs—No input connections to BEMC are normally required for exhaust systems except when exhaust capacity is over 10,000 CFM or exhaust system is of a critical nature, in which cases a status input (binary) should be provided. Status should be provided by sensing minimum static pressure or air flow.

Outputs—Provide one output per exhaust fan (2HP and larger only) to shut down and restart fan.

C.   *MECHANICAL COOLING SYSTEMS:*

Chilled Water Systems

Inputs—Provide an analog temperature sensor input for the chilled water supply and an analog temperature sensor input for the chilled water return. For water-cooled systems (cooling tower) provide an analog temperature sensor input for the condenser water leaving the chiller condenser.

Outputs—Chilled water cooling systems normally have compressor, chilled water pump and condenser water pump interlocks. The BEMC relay (output) should enable the critical component in the interlock scheme; generally this will be the chilled water pump. The compressor will automatically start up through a flow switch or pressure switch whenever the chilled water pump is operating. If the condenser water pump is not interlocked, the BEMC output should enable this pump as well. The tower fan is cycled by a native temperature control and requires no output connection. The BEMC output shall only enable and disable the normal run mode of the chilled water system and will have no other control function. The existing native controls will continue to operate the System whenever the BEMC output is enabled.

Direct Expansion Systems

Inputs—For systems 25 tons and larger, provide an analog temperature sensor input on the suction line at the compressor and an analog temperature sensor input on the liquid line from the condenser. For systems between 10 and 25 tons, provide only the liquid line temperature sensor. These sensors should be strap-on bulb type and should be tightly fastened to the refrigerant pipe, covered with thermal lubricant and the entire pipe and sensor assembly taped and insulated. For systems smaller than 10 tons, inputs are required.

Outputs—An independent condensing unit Enable/Disable

function relay connection to the BEMC is to be provided to allow the BEMC System to disable mechanical cooling operation under high demand or low ambient conditions, while still allowing the AHU fan to operate. The BEMC interrupt relay should be installed in the native compressor run control circuit in such a way as to simulate a normal cooling satisfied system shutdown.

This can be done by causing the liquid line solenoid valve to close on larger systems or by interrupting the compressor start control circuit on smaller systems. Interrupting the thermostat cooling wire will not prevent the compressor from restarting and is not acceptable.

D.   HEATING SYSTEMS

Boilers (all fuel sources)

Inputs—On hot water boilers over 1 million Btus, an analog temperature sensor should be mounted in the water vessel or in the hot water pipe immediately out of the boiler. A strapon bulb sensor may be used if it is fastened securely to the bare pipe and well wrapped with insulation. On steam boilers over 3 million Btus, an analog steam pressure sensor shall be installed in the steam header off of the boiler steam outlet.

Outputs—Steam and hot water boiler firing must always be under the full control of the native boiler controls and safeties and the BEMC output connection will enable and disable the local control sequence. Seasonal shutdown and startup will be done manually by trained personnel.

Hot Water Heating Pumps

Inputs—No inputs to BEMC are required for heating pumps.

Outputs—Provide an interrupt relay connection from the BEMC to the pump starter. This output shall be capable of turning off the heating pump based on either the outdoor temperature input or an average of indoor temperature inputs but will not be time scheduled on/off.

E. *LIGHTING SYSTEMS*

Inputs—A single analog ambient lighting sensor is required to provide switching level setpoints for all assigned lighting loads. This will include at least 50% general and pedestrian area lighting circuits in areas with good natural light from clerestory windows or skylights, and all outdor lighting, security lighting and parking lot lighting.

General Indoor Lighting

Inputs–Provide at least two BEMC output connections for all general area indoor lighting circuits: one output to switch to approximately 50% lighting level for janitorial servicing of the building and the second to switch to full lighting level for occupied hours. Unless already existing, the BEMC contractor will provide a magnetic contactor in an approved panel at or near each existing lighting panel location. Where several circuits of the same lighting function groups are located in the same breaker panel, a single multi-pole contactor may be used to interrupt these circuits.

Indoor Lighting, Daylighting

Outputs–Provide at least one BEMC output connection to control lighting circuits at or near skylights and clerestory fenestration areas. Unless already existing, the BEMC contractor will also provide a magnet contactor in an approved panel at or near each respective existing light panel. Where several circuits of the same lighting function group are located in the same breaker panel, a single multi-pole contactor may be used to interrupt these circuits.

Outdoor Lighting, Decorative and Security

Outputs–Provide at least one BEMC interrupt relay for each of the outdoor security lighting function groups. Also provide required lighting contactors if none currently exist. The BEMC relay is to be wired in series with any existing time clock contacts. Existing time clocks will be left operable, but all pins are to be removed.

Parking Lot Lighting

Outputs–Each parking lot lighting group will require a BEMC interrupt relay and a lighting contactor if none presently exist. The BEMC relay is to be wired in series with any existing time clock contacts. The existing time clocks will be left operable, but all pins are to be removed.

General Facility Sensing and Monitoring

These devices represent the building as a whole and are required for the application programs. They include outdoor conditions and utility meter tie-in. It is preferred that the utility provide a pulse to the computer; however, separate potential and current

General Facility Sensing and Monitoring

In addition to the ambient light sensor described previously, there are certain other inputs required to be provided to the BEMC system which are not directly associated with a single piece of equipment or system. These include:

transformers with a kWh transducer are also acceptable.

- a single global analog outdoor temperature input
- an ambient relative humidity input
- a digital or analog input which can be interpreted as building electrical demand (instantaneous)
- a digital or analog input which can be interpreted as facility electricity demand (accumulative).

Note: The electricity units to be reported and accumulated from this input by the BEMC system shall match the units as billed by the local electric utility for demand and consumption, i.e., either kW or kVA for demand and kWH for consumption. It is not required to read out or accumulate power factor and kVAH even when such units are included in the electrical billing calculation.

3.6 INSTALLATION

All work shall be included for a complete Energy Management system from remote sensors to data gathering panels to CPU to Output devices and all other necessary equipment to make the system operational. It will be the EMCS vendor's responsibility to monitor and control all points necessary to carry out the intent and details of this specification whether specifically listed or not. Also included are floor plans which show the locations of the mechanical equipment rooms. Detailed installation specifications follow.

Conductors—Control conductors provided by the contractor for connection from DGPs to equipment shall be copper with the type of insulation required to achieve the specified system reliability; however, they shall not be of lesser quality than type THW, or RHW. Conductor size and conduit fill shall be in accordance with the requirements of the National Electrical Code. Individual conductors installed in conduits shall be No. 14 AWG or larger.

Data transmission cables shall be run in conduit or underground duct.

ELECTRICAL CONDUIT SYSTEMS

Conduits shall be sized in accordance with the requirements of the National Electrical Code. Minimum size of conduits shall be 3/4 inch, except those used for exposed control and monitoring circuits may be 1/2 inch.

Changes in direction of runs shall be made with symmetrical bends or cast-metal fittings. Field-made bends and offsets shall be made with an approved hickey or conduit-bending machine. Crushed or deformed raceways shall not be installed.

Exposed conduits shall be installed parallel to, or at right angles with, the lines of the buildings or structures. Conduits shall be securely supported and fastened in place at intervals of not more than 10 feet with pipe straps, wall brackets, hangers, or ceiling trapeze. In no event shall a section of conduit more than 10 feet long be left unsupported. Horizontal conduit runs having bends shall be supported at the bends or immediately adjacent to the bend. Fastenings shall be by wood screws or screw-type nails to wood; by toggle bolts on hollow masonry units; by expansion bolts on concrete or brick; by machine screws, welded threaded studs, or spring-tension clamps on steel work. Nail-type nylon anchors or threaded studs driven in by powder charge and provided with lock washers and nuts may be used in lieu of expansion bolts, machine or wood screws. Threaded C-clamps shall not be used.

Raceways or pipe straps shall not be welded to steel structures. Holes cut to a depth of more than 1½ inch in reinforced concrete beams or to a depth of more than ¾ inch in concrete joists shall avoid cutting the main reinforcing bars. Holes not used shall be filled. In partitions of light steel construction, steel-metal screws may be used, and bar hangers may be attached with saddle ties of not less than No. 16 AWG double-strand zinc-coated steel wire. Conduits shall be fastened to all sheetmetal boxes and cabinets with two locknuts and a bushing. Bushings shall be installed on the ends of all conduits and shall be of the insulating type where required by the National Electrical Code.

Electric metallic tubing may only be installed in dry locations inside the buildings.

Flexible conduits of 12-inch minimum length shall be used for connections to motors and to equipment subject to vibration or movement, unless otherwise shown. Liquid tight flexible conduits shall be used in wet locations.

Wherever possible, EMCS conductors, conduits and pneumatic tubing shall not be routed through normally occupied, finished rooms, or on finished surfaces. Where it is necessary to install sensing devices on finished surfaces, the data transmission conductors shall be concealed in the most direct route to a point of concealment (such as above a ceiling), where it shall be routed via conduit to its DGP. Exposed EMS conduits, Wiremold, or tubing shall be painted by the EMCS contractor to match the existing spaces or surfaces to which it is attached.

SENSORS AND CONTROL DEVICES

Installation of sensors and control devices on existing systems shall be accomplished with minimum disruption to normally operating systems. When shut-down of operating equipment and controls cannot be reasonably avoided for the installation of sensors or control devices, the contractor must obtain approval prior to starting work.

To minimize downtime of any system, the contractor shall have prefabricated the item to be installed to the maximum extent possible, and shall have all necessary materials and personnel and return the system or equipment to normal operation as soon as possible.

DUCT MOUNTED SENSING DEVICES

Except as specified above for systems containing redundant air handling equipment, installation of sensors in air ducts shall be accomplished without shutdown of the air handling system. Where dry bulb or other type sensors penetrate the duct, suitable grommets shall be installed by the contractor through the duct wall to seal and retain duct insulation or liner material, to retain the integrity of vapor seals, and to form a tight seal between the sensor and the duct. The installation of wet bulb (or humidity) sensors shall also include contractor-installed hinged access panel, sized to suit necessary access to the sensor in the duct.

PART III. ALTERNATES

ALTERNATE #1

A system shall be provided to allow the development of customized Management reports. This shall be accomplished by utilization of a Personal Computer and all support software and hardware. English language message capability shall be provided along with access to all input/output and data stored in memory. The terminal shall be located in the Energy Engineer's office or equipped for remote communication via modem (300/1200 Baud).

ALTERNATE #2

Provide a maintenance contract for the CPU, operators' terminals and data gathering panels. The yearly cost for each location shall be specified. The maintenance contract shall provide all equipment, maintenance, troubleshooting and repair of the equipment listed above to keep the EMCS operational. No maintenance shall be provided on the sensors or wiring to DGPs.

ALTERNATE #3

Same as Alternate #2 except include all sensors and wiring.

ALTERNATE #4

The EMCS shall be capable of being expanded to include fire alarm and security access. This shall be accomplished by the addition of another operator's terminal, data gathering panels and sensors with wiring only. The CPU and DGPs shall meet applicable fire codes, NFPA-72D Type I, and be UL listed.

SPECIFICATION 3 — HVAC and Lighting Control System Performance Specification
(Provided by George R. Owens, *P.E., C.E.M.*)

This deceptively simple guide can be used only when the issuing party has complete technical competence in every aspect of EMCS design, installation, operation and maintenance.

The specification is for a full-function mid-range energy management and control system, handling 500,000 sq ft of space. The system has 128 points of control including HVAC, lighting, and base sensors. It had a one-year payback.

I.   GENERAL

This performance specification is for a lighting and energy control system. The control system shall be centrally located near the management office and shall provide automatic controls for the lighting and energy systems as outlined below. The price shall include engineering, preparation of drawings, all hardware, wiring, conduit, etc., the energy control system and interfaces and installation and start-up supervision. All equipment shall be Underwriters' Listed; all installation shall conform to any applicable national, state or local codes.

II. LIGHTING

   A. *Controller*—The controller shall be an electronic programmable time clock capable of time control of eight minimum lighting control schedules. The controller shall be all solid-state by design, shall have digital readout indicating the date and time and program; it shall have an integral or detachable keyboard for data and program entry; it shall have minimum 24-hour battery backup of the program; holidays, leap year shall be preprogrammable; multiple on-off periods for each channel shall be user selectable in one minute increments to be programmed on a daily basis. A total of eight days per point shall be included. Operator, programmer, installation, and maintenance manuals shall be provided to the owner.

   B. All relays, contactors and other equipment required to interface to the existing lighting circuits shall be provided by the vendor.

   C. *Wiring and Conduit*—All wire, conduit and installation labor shall be provided by the contractor.

   D. *Control Points*—The following control points are required to be included separately under this proposal:
     1. Parking lot lights,
     2. Entrance lights,
     3. High bay down lights,
     4. Decorative lights,
     5. Recreation area lights,
     6. Service corridor lighting.

Photocell control shall be maintained for all lighting presently containing photocells. Owner has right to review and change the designation of specific lighting types and the number of circuits up to a maximum of eight prior to installation.

III. HVAC ENERGY CONTROL

   A. *Controller*—The energy management controller shall have capabilities of performing energy management functions as listed below:

1. Analog temperature indication,
2. Electric utility meter interface,
3. Scheduled start-stop of HVAC units,
4. Optimum start-stop of HVAC units,
5. Duty cycle based upon analog temperature compensation,
6. Demand shedding based upon analog temperature compensation, and utility meter data,
7. Night setback control

The controller shall also have hard copy printout with user selectable logs to determine proper operation of the system and provide hard copy data. The alarms for problems shall be indicated by means of an audible horn and lights and/or CRT screen display.

B. *Loads*—Each unit over 50 tons shall have the fan controlled, two stages of air conditioning compressors controlled and one stage of electric heat controlled. All units below 50 tons shall have 2 points of control. These units may be grouped by function up to 4 units per group if below 25 tons. One analog input is required for each HVAC group. Outdoor analog temperature input is also required.

14

EMC System Procurement, Installation, Fine-Tuning and Maintenance

Once bid documents have been prepared, the remaining steps relate primarily to procuring the system and working with the contractor during installation, testing and fine-tuning the system, and maintenance, as follows.

OBTAINING BIDS

At least three competitive bids should be obtained.

Because manufacturers' names for various system elements differ, it is suggested that all bidders should be required to utilize generic terminology, as defined in the specifications. Manufacturers may, at their option, provide a glossary of terms which identifies the trade names used for the generically described devices involved. Prospective providers also should be required to indicate those elements of plans and specs which they are for whatever reason unable to comply with, alternatives available to provide a similar function, and alternatives and options which may not have been considered along with the cost of the hardware and software, and other alternatives.

CONTRACTOR SELECTION

Proposals received from the various contractors bidding on system installation should be reviewed carefully. Three separate reviews are involved, as follows.

First, all proposals should be reviewed independently by the chief operating engineer and selected members of his staff. The purpose of this review is two-fold. It enables a technical evaluation, and it involves engineering staff substantially in the process which eventually will result in a substantial change for them.

Second, all proposals should be reviewed independently by the EMCS consultant.

Third, the chief operating engineer and the EMCS consultant, and their respective support personnel, if any, should meet together to discuss their findings, along with one or more members of building management.

Contract award should *not* be based on low bid only. Rather, it should be based on *value*, implying an evaluation of the cost-benefits of each proposed system. For example, if a system offered does not adhere fully to plans and specs, but would deliver additional benefits, the value of these benefits must be considered. In this regard, the first question to ask is, "Are the additional benefits worthwhile?" If the additional functions are not really needed, then their benefit cannot be given high marks for the facility involved.

In general, it is suggested that a value ranking methodology be established, and that each item of concern be given a certain point total, the overall total adding up to 100 points. Thus, a value rating system may look somewhat like this:

Factors	Weights
Compliance with Requirements	10
Communications Standardization	9
Effective Description of Device and Components	5
Vendor Reputation	5
Training Support	9
Field Maintenance Support Quality	7
Ease of System Expansion	7
Fire/Security System Integration Capability	6
Available Options	5
Prior EMC System Compatibility	7
Vendor Experience and Stability	4
Adequacy of Time Schedule	5
Hardware and Software Quality	5
Installation Costs	6

Factors	Weights
Maintenance Costs	6
Guarantees and Warranties	4
Total	100

Certain of these factors can be evaluated only on the basis of the contractor's proposal, for example, warranties and guarantees, adequacy of software and hardware, etc. But other factors can be determined also by certain basic investigations which should be performed, and for which contractors should be required to provide information. Typical information includes the names and addresses of others for whom the contractor has performed work, as well as the names and addresses of persons responsible for facilities which have installed similar systems using some or all of the same components (regardless of the contractor involved). Contacting these individuals will provide input with regard to the reliability of the systems and components, the attitudes and actual capabilities of the contractor, etc.

Note: When others are being contacted, it is extremely important to ask the right questions. Past experience indicates that the questions asked have been too general, and that some of those answering the questions really were not in a position to provide fully accurate responses. As such, it is suggested that the EMC system consultant should develop a list of questions for which responses are needed. When two or three offerors are identified as the ones most likely to receive the award, it may even be desirable to have the EMC system consultant investigate the installation in person.

If this value ranking approach is used, it should be devised by the EMCS consultant and members of facility engineering staff working together. This should be performed subsequent to preparation of plans and specifications and prior to announcement of the request for proposal. The bidder evaluation form should be included with plans and specifications.

The value rankings achieved should then be evaluated in terms of all cost factors involved. In this way, certain other factors will come to light. For example, a system which is capable of integrating existing EMC systems and devices will probably cost less than one which requires new equipment.

Clearly, the process of evaluating a proposal and selecting a contractor is not simple. It requires extensive review and analysis of numerous factors. This is fundamentally necessary, however, since the system is likely to represent a substantial sum, and must be able to support the function of the building.

POST AWARD REVIEW

Once the contractor has been selected and funds have been appropriated for the project, the contractor should submit a project management plan (within 30 days after award of contract). The project management plan should consist of at least the following factors:

1. Organizational chart, indicating project team members, i.e., project manager, field superintendent, foreman, etc.

2. Program plan including a statement of tasks.

3. Project schedule.

4. Reporting format.

5. Submittal procedure, and

6. program meeting arrangements.

Once a project management plan has been approved, a series of meetings should be held to help ensure full understanding of all key persons' concerns. By taking these steps, many potential problems can be minimized, if not avoided altogether.

For this reason, it is essential to establish definite ground rules and working guidelines between the contractor, the building owner, and the EMC system consultant. The contractor must understand established submittal procedures, time frames for completion of various stages of the project, and other contractual requirements.

Authority for making monetary and other decisions must be clearly assigned to specific people. The contractor's representative must be known and accessible. The facility representative with the responsibility and power to make commitments and decisions must be identified and be readily available to avoid delays and expedite completion of the job.

One of the best ways to keep the communication flowing between the construction supervisor, support people, inspectors, and others is a daily written progress report. Such reports can be very effective, providing they address real concerns: What was accomplished? What problems surfaced today which could cause problems later? Can the schedule still be met? Etc.

Three specific meetings should be held to expedite progress and minimize problems.

Preliminary Design Review

The purpose of the preliminary design review is to clear up major inconsistencies, and to authorize the contractor to start ordering certain basic pieces of equipment and materials.

The meeting will last for one or two days, involving contractors and major subcontractor personnel, facility engineering staff (or chief operating engineer only), the EMC system consultant, and if desired, representative(s) of company administration.

The matters of primary concern include:

1. Review and resolution of any misunderstandings.

2. Review and resolution of minor inconsistencies between the proposal and plans and specifications.

3. Identification of other matters which require modification, review, etc., at subsequent meetings.

4. Establishment of schedule for conduct of remaining post-award, pre-installation activities, such as meetings.

5. Approval for the contractor to order specified pieces of equipment or materials requiring long lead times.

Minutes or an actual transcript of the meeting should be made to help ensure that all persons attending the meeting, and perhaps others, can quickly refresh their memories with regard to who is responsible for what.

Second Review

The second review meeting, involving the same people as the first, perhaps supplemented by others, if necessary, should be used to:

1. Review and approve contractor-suggested methods for resolving inconsistencies discussed in the first meeting.

2. Identify, review, and resolve any additional inconsistencies.

3. Finalize design and specifications.

4. Finalize test requirements and procedures.

5. Identify the names of persons who will be responsible for implementing installation, including those representing the contractor, major subcontractors, if any, operating engineering staff, and the EMC system consultant.

6. Create a detailed draft timetable, identifying specifically when certain functions are to be started and completed.

Third Review

Once the timetable has been prepared, it should be circulated to all concerned parties for review. Comments regarding likely conflicts, items for which more time is needed, etc., should be made in writing, and addressed to a specified individual, such as the chief operating engineer or the EMC system consultant.

Following receipt of materials, by a specified cut-off date, a third review should be held. This would involve a meeting including those who attended the first two, as well as additional persons who will be substantially involved in system installation. This will give the people who will be working together an opportunity to meet one another, discuss communications techniques, and such other matters whose clarification will result in far fewer problems later on.

In essence, installation should not commence until all matters of concern are resolved; a timetable has been established and agreed upon, and those who will be involved in system installation have had opportunity to familiarize themselves with the various techniques and procedures that will be used.

SHOP DRAWING REVIEW

One of the most important tasks during the construction phase is shop drawing review. This is the point at which the contractor's intentions are made clear. Ideally, there will be no exceptions to project specifications. Where there is doubt, as about the level of quality intended, the contractor will offer a product which exceeds the minimum requirements.

Read all of the fine print. Failure to do so can result in serious problems later on. Although the shop drawing reviewer is not legally responsible for deficiencies missed, it is far better to discover potential problems at the shop drawing phase before the work is installed. Fine-print items, such as environmental requirements of FIDs, electrical characteristics of central equipment, lease telephone line requirements or requirements for accessories that are not standard but necessary, can be easily missed and have serious implications in the performance of the system.

ON-SITE DEBUGGING

After the system has been installed, but before acceptance tests are begun, the contractor should operate and calibrate each point in the entire system. Personnel should be located at both central and remote locations. In that way, when a certain function is performed, values are read at the central station, and the actual function at the remote station is observed and readings are verified. For example, if a temperature of 70F is called for and registered at the central station, it should also be verified by human observation at the location specified. If a tolerance of as small as ±0.5F is required and specified, the consultant should make sure that the instruments used to confirm the signals being sent back to the central component actually operate within this tolerance.

In checking and calibrating each point of the system, the contractor should submit a written report of the procedures and results. Details such as the time of day, what specifically was tested, what happened the first time through, whether the point was recalibrated, what procedures were performed to establish that the individual point was actually operating, etc.—all should be supplied.

The EMCS processors can be used for logging purposes as an aid to the debugging procedure. Alarm and logging procedures could be established which, together with the day-to-day operations, allow for the processors to track and flag potential problem areas. The printers can produce hard-copy records of these logs that are extremely useful for detecting possible problem areas.

Data log printouts, alarm printouts, or CRT readings should not be relied on heavily as indications that things are working smoothly. The possibility exists that erroneous signals may be sent back from the remote sensors indicating proper working of the equipment when in fact the equipment may be malfunctioning. Once again, human verification of the printer or CRT readings must be obtained to assure that the sensors are in working order. This should be done for all points in all locations being checked in the random-checking procedure.

The number of points to be checked in this debugging procedure will vary with the particular system involved.

After calibrating each individual point in the EMC system and ascertaining that they are all in operating order, the next step is to operate the system as a complete entity, for acceptance testing.

FINE-TUNING AND
LONG-TERM CONSIDERATIONS

The fine-tuning process is one in which operators are given an opportunity to experiment with the system, and to correct unforeseen difficulties. The fine-tuning period should encompass one year from the date on which the system is accepted. During that time, contractor personnel, building engineering personnel, system operation personnel, and the EMC system consultant should meet on a monthly basis to review the results of experimentation, unforeseen problems, and related matters.

In order for energy managers to obtain maximum efficiency from the system, they will attempt various control strategies. They may wish to establish trend logs for certain buildings or try different strategies for different components. Things can go wrong during these experiments. For example, temperatures may be too hot in one building and too cold in another. See Chapter 17 for a detailed discussion of these issues.

Cooperation must be gained between energy managers and the people who work in the controlled buildings. The people working in these buildings should be notified that the system is being fine-tuned, and should be requested to leave the thermostats alone, not plug up ducts, or do anything to jeopardize the success of the operation. They must be able to realize that their cooperation is essential, and that a little discomfort during the fine-tuning process is a small price to pay for a successful EMC system.

Fine-tuning also will reveal things that cannot be foreseen during the system design. Calculations can show that a temperature of 78F in a certain building will save 10,000 Btu, but the theoretical parameters on which those calculations are based may have to be continually modified during actual practice. For example, the designer may determine that shutting off a chiller in a certain building at 4 p.m. and allowing the chiller's flywheel effect to keep the building cool until 5 p.m. will save energy.

As it turns out, there is an unforeseen requirement that a large door to the outside must remain open during that hour, thus requiring the chiller to remain in operation. Obviously, such problems cannot all be predicted, nor will the system do 100% of what it was designed to do on the very first day.

SYSTEM DOCUMENTATION

As outlined in the specification, as-built documentation must be provided within a specified period after system installation and start-up, but prior to system acceptance. The reason for this is to ensure that the somewhat tedious item for the EMCS installer is provided before release of final payment or retainage. It is critical that these documents include all site modifications which are not reflected on the shop drawings. In addition, System Documentation should include a full range of information regarding system applications and hardware.

As a general rule, the areas outlined below should be included with documentation, and a final step in the system acceptance procedure is to verify the accuracy of the submittal.

1. System Summary: Includes names and contact points for all key project participants, manufacturer of EMCS, start-up date, and a brief outline of any issues requiring owner attention.

2. System Architecture: Manufacturer, model, software/hardware address and physical wiring architecture for all major system components, system point list, and details regarding actual location of modules.

3. System Software: A floppy disk or hard copy of complete system program instructions including algorithms (where appropriate), parameters, setpoints and time-of-day schedules should be provided at start-up.

4. Ancillary Equipment: Manufacturer, model, and location of all relays, contactors, transducers, etc., and the controlled load or piece of system architecture with which they are associated.

5. Applications Data: Any modifications to control wiring, electrical systems or existing facility equipment to accommodate the EMCS must be documented to facilitate long-term maintenance.

6. Shop Drawings: Modified shop drawings reflecting actual as-built conditions for all system interface.

It is helpful for some building owners to receive multiple copies of these documents to be kept on-site, and in the owner's office. Another useful tool may be to receive a version of these documents in 8½ x 11 inch or 11 x 17 inch format. This copy may be placed in a three-ring binder and function as a convenient desk reference during EMCS program management.

SYSTEM MAINTENANCE

Once the EMC system is functioning as designed, it is important to implement an effective program management and preventive maintenance program. Chapter 17 on Program Management covers this topic in greater detail. Without an ongoing equipment maintenance program coupled with solid EMCS program management, problems will occur and people will lose confidence in the system.

The central processor, the computer, CRTs, printers, field devices and all the other electronic equipment must be maintained. Diagnostic programs for the various computer components should be required as part of the specifications. Wiring diagrams of the system also are essential during maintenance procedures; someone who understands the diagrams must be available.

In deciding whether to use in-house maintenance or an outside source, several considerations must be weighed. Much depends on the size of the operation. If a large in-house maintenance staff already exists, appropriate personnel may be able to take care of electronic gear. Recognize, however, that spare parts must be stocked, space to stock those parts must be allocated, and personnel to keep inventory current must be assigned. Large facilities may not consider this an additional burden, but smaller ones may not wish to be bothered with additional people, parts, etc. For smaller facilities, therefore, a maintenance contract with an outside service organization may be the answer.

If an outside maintenance organization is to be used, competitive bids should be solicited. All documentation mentioned in the specifications must be available to the bidders. If the specifications do not require wiring diagrams of the system to be delivered, the only feasible maintenance organization is the supplier of the equipment. Obtaining all necessary documentation is a requirement that should not be overlooked; otherwise the bidding may not be competitive at all.

Once the decision is made, it is beneficial to have the maintenance staff actively engaged in the debugging procedure. They can participate with the operators in the initial training provided by the contractor. They will be familiar with the debugging process, and will have lived with the system, thereby gaining experience so that there will be few surprises later if problems arise.

Note that the *complete system requires maintenance,* including the computer, peripheral equipment, data links, local sensors and controllers, trunk wiring, etc.

The person responsible for maintenance must make sure that the total system is operating and on line. Monthly diagnostic programs on printers, CRTs, etc., are not enough. End point sensors must also receive continuous attention as should FIDs and other electronic equipment located at remote facilities.

15

Tips on EMCS Specification, Vendor Selection, Operation

George R. Owens, *P.E., C.E.M*

This chapter describes personal experiences in the proper specification and selection of an EMC system from an optimum cost/benefit standpoint; the installation and programming of the EMC system; and the keys to the continuing long-term success of the installation.

A preprequisite to obtaining a successful EMC system installation is a thorough knowledge of the current state of the art in energy management software and hardware. If you do not have sufficient knowledge, you must either develop it or hire it. The education process should require from four to six months of researching the topic. This time is well spent, considering that an EMC system may cost between one quarter and two to three million dollars and be required to be in service for ten to twenty years or longer. With this magnitude of importance, six months spent prior to purchase is superior to five years of costly correction of mistakes or misunderstandings later.

SPECIFICATIONS

A good specification is the single most important item to obtain an EMC system that will realize energy conservation savings at the least installed cost. The specification has to be detailed enough to pin down exactly what a vendor is expected to provide. For example, a detailed point list must be provided with a specification to insure all the points required are provided. Conversely, the specification must be general enough to allow for the differences in vendor's hardware. For instance, communication techniques may be technically different yet perform the same function and should be treated equivalently.

Finally, a specification isn't worth the paper it's printed on if it isn't adhered to. If there are changes after the specification is issued prior to bidding, these changes should be issued in writing to all bidders.

Software

Before beginning to write a specification for an EMC system, determine the energy management control strategies that are applicable to the building(s) in question. Without extensive software that can be applied to your energy consuming devices, you'll end up with a $1,000,000 pile of transistors and wires which makes a fine time clock and not much else. Normally the energy conservation calculations required for justification will indicate what control strategies are necessary. In any case, determine each strategy required, and include a description in the body of the specification.

In writing the specification make sure that the full cost of software must be included in the vendor's quotation. If software costs are not included in the quotation, you will be in for a costly surprise if a vendor approaches you for an add-on later as software costs can run 10–20% of the total job.

Two other areas of software concern are the method for entering user data (i.e. temperature limits, start/stop times, adding points), and user generated custom programming capabilities. Both of these functions must be available to the user and should be accomplished while the system is running the existing equipment and programs. Without these features, the flexibility of the EMC system will be severely limited and the delay and cost to make additions or changes can be prohibitive.

The system will be changed—must be readily changeable—to optimize present savings and accommodate future innovations.

POINT SELECTION

Since a specification is the single most important item to obtain an optimum EMC system, the points chosen to monitor and control are the most important cost control device available to the system designer. If not enough points are included, the energy management strategies specified will not be achieved. On the other hand, if unnecessary or "nice to have" points are included, the cost of the system will increase significantly. Table 15-1 lists the majority of the points available to the user and is broken down into the following three categories.

Prior to selecting points for a large device, such as a chiller or boiler, contact the vendor about control strategies and/or warranty implications. Potential problems can be avoided if these checks are made prior to point selection.

Necessary

A. These devices cannot be deleted without compromising the value of the EMC system.

Justify

B. These devices are not absolutely necessary to the success of the EMC system. However, they can speed information gathering and troubleshooting of the EMC system and the controlled equipment. Each of these devices must be justified individually and fit into the budget.

Unnecessary

C. These devices significantly increase costs of the EMC system without improving payback. Any in the category must be questioned strongly and eliminated (unless a specific cause required them) to obtain a cost-effective system.

Table 15-1. Point Selection

Controlled Device	Necessary Points	Points To Be Justified	Unnecessary Points
Environmental Sensors	A. 1 indoor dry bulb for monitoring space conditions and operation of cooling and heating. B. 1 outdoor dry bulb for operation of cooling, heating and freeze protection. C. 1 outdoor dew point or relative humidity for enthalpy program.	A. 1 or 2 indoor dry bulb sensor per floor—if building is large or conditions from floor to floor vary.	A. 1 indoor dry bulb per zone —rarely justified. Local loop control is more cost effective.
Energy Consumption	Transducer from utilities' meter is most accurate. Watt transducer if utility signal not available.	Submetering of partial loads, separate buildings and large equipment may be justified to optimize energy consumption.	
Fan Units	A. Start/Stop of each supply fan. B. Flow verification of each fan operating. C. Damper control at each fan. D. Return air dry bulb and dew point (or relative humidity) used for enthalpy.	A. Start/Stop of return fans separately only if operated other than supply fans (i.e., duty cycle supply but not return fans). C. Damper verification of operation can be justified in some cases. D. Temperatures at each fan such as mixed air and supply air will aid in troubleshooting problems.	E. Not usually justified. A. Filter status—maintenance inspection is adequate. B. Vibration—Local switches for shut down are better.

Air-Conditioning Units	A. Start/Stop of air conditioning unit. B. Verification of motor running.	C. On large units, the following is probably justified. Items 1 & 2 are used in calculating efficiency. 1. Chilled water temperature supply and return. 2. Chiller operating amps. 3. Alarm if one of the safeties is tripped.
Boilers	A. Start/Stop of boilers and hot water circulating pumps. B. Verification of motor running.	C. Hot water temperature for setback adjustment and verification of boiler operation in cold weather (freeze protection).

CONFIGURATION AND HARDWARE

Only after the software and the control points are selected, should you begin to consider the EMC system configuration and the hardware. Unfortunately, the greatest portion of time is expended in specifying the hardware in detail when the time could be better spent refining the software or optimizing the point selection.

Hardware specifications should be performance specifications, not detailed specifications. This will allow for differences between vendors yet produce a system that will meet all expectations at the optimum cost.

A. Central Processor Unit (CPU)

In this section reference to the CPU applies to both microprocessor- and minicomputer-based systems. The CPU is the "brain" in both systems and handles all of the input/output of data, software storage, calculations and control actions. The two key parameters of the CPU are the execution speed and the amount of memory as defined in EMC system terms, not computer terms. Important is the response time to alarm conditions, the time for a control action after a signal is issued and how often analog signals are updated in the CPU's memory. The only way to be sure that sufficient memory is installed is to specify the size of the software and the number of points. For example the specification might read:

The CPU shall be capable of handling the following as a minimum:

1. 2 buildings
2. 5 utility meters
3. 100 points

In addition, the CPU shall be expanded by the addition of circuit boards to handle the following:

1. 10 buildings
2. 20 utility meters
3. 2000 points

Total number of points are to be specified in the point schedule (see Chapter 9 for examples).

An EMCS shall be capable of the following:

1. Start/Stop programs
2. Night Setback/Setup
3. Optimum Start/Stop program
4. Demand program
5. Duty cycle program
6. 20 custom routines

B. Personal Computers, Terminals and Printers

Specify Personal Computers (PCs) as necessary for local and/or remote communications. Though PCs are rapidly becoming the interface device of choice for most EMCS users, it is also important to specify enough local terminals for operator interaction. PCs, terminals and printers are expensive and impact the bottom line job cost, so proceed with caution. Local and remote interaction should be via PC with English language prompts. However, simple terminal interaction may be useful for interrogation and diagnosis. Printers are useful for hard copy of PC-generated information, and as standalone devices for alarm reporting. Printers should be wide carriage (8½" paper minimum) and have English language descriptors and bold face alarm messages.

C. Field Devices

Sensors shall be ±1% accuracy or ±1°F, ±1 PSI, etc., whichever is smaller.

D. Wiring

Two options exist:

1. Hard wiring to either the CPU (small systems) or local multiplex points.

2. Pull multiplex where the same wire carries data to the remote devices. Investigate for cost savings.

MISCELLANEOUS

The specification should include a discussion of the following topics as they apply to the system being designed:

A. Battery backup of the CPU in case of power outage. Minimum —4 hrs.

B. Software and user data file storage for system re-initialization after shutdown (i.e., cassette tape, floppy disk or hard disk).

C. Installation details such as conduit, exposed wiring, sensor mounting, etc.

D. Warranty—1 year.

E. Insurance requirements of your company's insurer.

F. Installation schedule requirements.

SELECTING A VENDOR

Remember, once you select a vendor, consider yourself married to that vendor for many years. Like it or not, vendors' equipment is not interchangeable. Changing vendors after the EMCS is installed is extremely expensive. Before you choose a vendor, make sure you choose wisely with a long-term relationship in mind.

The selection procedure occurs after the receipt of all quotations from the prospective vendors. The phases of selecting the best EMC system are:

A. The quotations are reviewed to insure that all requested supporting documents, descriptions and drawings have been supplied.

B. The proposals are reviewed for technical content. Any questions are presented to the vendor for their reply. At this point any vendor that does not meet the minimum requirements is dropped.

C. The installed system price is reviewed with each vendor to insure that *all* items are covered by the quote. Usually, three ranges of quotes will be received. Any quotation that is 25% above the mid-range can be eliminated. The quotations that are 25% below the mid-range may represent an opportunity to reduce the cost of the system. However, each of these proposals must be re-reviewed carefully to insure the EMCS shall perform as anticipated. After this review, the selection should be narrowed down to two or three qualified vendors.

D. To make a proper decision about the final selection requires learning as much about the prospective EMCS as if you already owned it. As a minimum you must:

1. View at least one operating system similar to your application.

2. Discuss problems with at least two other EMCS owners of systems that are comparable to your own.

3. Review the vendor's engineering and maintenance staff at the vendor's office.

4. Understand how the software works and how to program the software. Also, have demonstrated to you, the addition and deletion of points, changing of Start/Stop times, custom programming and alarm messages.

Table 15-2 is a vendor selection comparison sheet that should be used to condense and tabulate the information required to make an EMCS decision.

INSTALLATION

The installation phase is the time to insure that you get what you pay for. If details are missed in this phase, the costs can be appreciable in energy waste, equipment damage and the cost to correct the oversight.

Before work commences, sit down with the vendor and develop a realistic schedule detailing when each phase of the work will occur.

Phase One is engineering where detailed construction drawings are prepared. All connections to existing equipment must be shown with all wires and tubing labeled. All conduit, wiring, equipment locations

Table 15-2. EMCS Vendor Evaluation

	Vendor No. 1	*Vendor No. 2*
Base Bid System Price	Include the full installed cost of the system.	
Service Contract Cost/Yr	Assumes a full service contract is obtained. Affects payback of the system, also what is covered and when.	
Service Reputation and size of local staff	Again, since the vendor's organization must be dealt with on a continuing basis, this is important.	
Total Number of Systems Installed	Both total number and the number identical to the specification. You may want to reconsider a vendor with only a few units installed.	
Maximum Size of Systems Quoted	For future expansion in number of locations, points and fire/security applications.	
Software	1. Does it meet the specification? 2. Additional features beyond this specification?	
Reports	How extensive is the management reporting function?	
Schedule	Installation and start-up dates.	
Engineering Support	Both design and trouble-shooting depth and location.	
Special Features	List here anything vendor can offer beyond specification.	
Comments	Additional space for comments.	
Conclusion	A brief paragraph here to describe the conclusions reached about the vendor's system with highlights of the best features.	

and tubing runs should be placed on a floor plan of the building. No construction shall proceed without owner approval of the engineering drawings.

Phase Two is the ordering of equipment by the vendor. This is the vendor's responsibility. However, review with the vendor the status of the ordering and equipment delivery to insure completion of the job on schedule.

Phase Three—During construction, review the job progress with the vendor. Tour the job site(s) to insure a quality installation. Insist upon correction of any discrepancies. Before any existing equipment is modified, the owner must be informed of the changes to allow orderly shutdown and start up. Modifications to existing equipment shall not violate the existing fire, freeze and interlock safeties. Local manual override shall be provided.

Phase Four—After construction prior to start up, the complete system must be checked for proper operation. The CPU and terminals must be operating. All the software strategies must be verified. The data panels must be tied into the CPU and proved to be communicating properly. Each sensor must be responding properly and calibrated in the field.

OPERATION AND MAINTENANCE

The single most important key to a successful ongoing energy management system is: one person must be assigned and held accountable for the day-to-day operation of the EMCS. The EMCS is going to have problems. It is not a question of "if" but a matter of "when." If these problems are not recognized and corrected, the EMC system performance will deteriorate to a state of shambles. *This point cannot be overemphasized.*

At least one person (the EMCS manager) must be knowledgeable enough about the EMCS to write programs, diagnose failures to the point that a determination can be made of whether it is an EMCS or equipment problem, and insure the continuing success of the EMCS. Training should be a combination of vendor-supplied training off-site and hands-on training on-site.

The operators of the system should have enough training to change operating schedules, respond to alarms, diagnose simple problems and most importantly, know when to ask for more help. Eventually one or more of the operators should be trained to make program changes and be a backup to the EMCS manager.

Personnel at the controlled location must be trained on the function of the EMCS and how to respond to local problems.

Programming the EMCS yourself is probably the single most effective training available. Most vendors include programming time in their pricing. However, by programming it yourself, you have complete control of the software and know how to specifically solve a problem.

Document your software. If changes are made to the EMCS software, make sure they are written down to allow troubleshooting of a problem.

A maintenance contract on the complete EMC system is strongly recommended. Since few building management organizations can support technicians qualified on the EMCS, this support is better left to the vendor. But before you sign the agreement, assure yourself that all items are covered (both parts and labor) on a twenty-four hour basis. Also determine a preventive maintenance schedule and insure the vendor follows it.

CONCLUSION

The chapter provides the procurer of an EMCS with guidelines for the proper specification, selection, operation and maintenance of an energy management system. The key factors outlined can mean the difference between a successful EMCS installation and a very expensive glorified time clock. Keep in mind that the EMCS is a tool for energy conservation, not the end result. By applying good energy engineering to the process of selection and operation of the EMCS, the energy savings predicted will be realized to the fullest extent possible.

16

Training the EMC System Staff

One of the major objectives of a training program is to develop a positive attitude among the personnel who will be involved with the system, since its success depends upon the attitudes of the personnel who will interact with it. It is essential that the staff assigned to operate the system become familiar with the hardware and operational format. Familiarity will minimize any threat that the system may present to the staff. Knowledge and understanding instill confidence and promote positive response.

It is suggested that the training program should be conducted over a period of time, such as six months, to help ensure that information imparted by training personnel is comprehended, to allow those trained to obtain experience, and to identify areas where more training and guidance may be needed. It is essential that the training program be geared to the functions performed by existing staff and the actual level of capability they possess. In some cases, a certain degree of remedial training may be required, depending on the education and experience of operating and maintenance personnel. In the case of programming personnel who will be hired by the facility, experience in programming a broad range of EMC systems is desirable. It must be a prerequisite for the programmer(s) to be fluent with the particular EMCS installed in the facility.

The training program outline shown in Table 16-1 is sophisticated. If in-house programming is not to be performed at first, certain elements of the training will be deferred, assuming that in-house programming will be performed later.

As shown in Table 16-1, the first phase of training should relate to the basics of the system, including proper operation of the CPU

Table 16-1. Training Program Outline

FIRST PHASE TRAINING

Operations, Maintenance, and Programming

* EMC system description
* Component function, interface and operation
* Logging mechanism and description
* Operation of computer and peripherals
* Operator control function
* Color graphic generation
* Troubleshooting EMCS components
* Preventive maintenance of all EMCS components
* Sensor and control maintenance and calibration
* Software description
* Application program
* Front End Monitoring and Programming for Parameters and Setpoints

SECOND AND THIRD PHASE

Operation

* Remote Communications and Program Management Functions
* Familiarity with HVAC systems
* Familiarity with software (application program)
* Operation control function
* Report generation
* Interfaces

Maintenance

* System and component function
* Spare parts inventory requirements
* Preventive maintenance of all EMCS components
* Troubleshooting
* Sensor and control maintenance and calibration
* System design and layout

Programming

* System design and layout
* Interruption logic
* Interfaces
* Software diagnostics
* Communications software
* Applications program manipulation and design
* File management
* Programming concepts

and peripheral devices, operating basis, elementary preventive maintenance, etc.

The second phase of training, conducted some two months after the first, would be geared toward operators, programmers (assuming they are used), and equipment maintenance personnel. Programmers would be given instruction in the computer language used and related concerns; equipment maintenance personnel would be instructed on troubleshooting, overall preventive maintenance, and sensor and control maintenance and calculation.

The third phase of training, conducted two months or so after the second, would relate primarily to operators and programmers. Subjects covered would include, among others, CPU architecture, interruption logic, interfaces, software diagnostics, communications software, file management, and advanced language and programming concepts.

The contractor should be in a position to provide additional training as it becomes necessary; for example, due to replacement of personnel or modification of equipment, on a separate contract basis.

17

EMCS Program Management

Facility managers are better equipped than ever before to manage successful EMCS programs. Optimum program performance is totally achievable in most installations. Over the past 10 years, managers have amassed a wealth of knowledge on this topic. More importantly, we have established the components that are essential to evaluating the EMCS, and implementing a successful program. That is the topic of this chapter, Program Management.

It is important to present, and validate, the practical rules-of-thumb and considerations which facility managers must bear in mind when dealing with computer control systems on a daily basis. However, this concept of program management goes beyond day-to-day operational considerations, because technology is not static. Facility managers find themselves constantly faced with new technology. An example is the microcomputer, which is more prevalent in the controls industry, and is being used quite successfully to enhance productivity.

The purpose of this chapter is to provide managers with data necessary to enhance EMCS programs. EMCS program management is treated under two headings: Components and Functions.

PROGRAM MANAGEMENT COMPONENTS

Energy managers are currently faced with a myriad of technical and logistical program management concerns. It is imperative for energy managers to reaffirm the importance of their programs, and the best sales tool available is a successful track record. To accomplish such a track record with EMS equipment requires an effective program designed to: (1) verify system operation, and (2) to access raw data and make computations in the process of fine-tuning system performance.

A primary focus must involve the use of personal computers (PC). In many cases automation is a major component of the global energy program, and one which requires diligence and time-intensive oversight, and front-end automation is essential.

The components of EMS program management which will be discussed are: (1) SYSTEM VERIFICATION via: Physical Maintenance, Security Issues, Performance and Management Reporting, and (2) SYSTEM FINE TUNING. It is also of prime importance to consider maintenance of controlled loads in the evaluation of EMS programs. Each of these concerns will be discussed in relation to program management.

SYSTEM VERIFICATION

Physical Maintenance

Maintenance requirements are typically quite limited with electronic equipment. As a result, in many cases long-term programs are not developed for addressing these needs. It is important to consider that all components of the EMS installation are not electronic. Normally, at least a portion of the project is comprised of off-the-shelf electrical gear, such as low-voltage relays and transformers of various sizes. This is particularly true where lighting and other ancillary equipment is to be controlled.

Redundancy has also been introduced into EMS control, especially on the low end of the spectrum where systems employ less than 100 points. Therefore, it is also an issue where control interfaces including Direct Digital Control (DDC) have interfaced with pneumatic controls, or other level controls. It is crucial to maintain the EMS interface to ensure that a failure does not occur which goes unnoticed due to this control redundancy.

The primary focus of physical maintenance is to ensure that EMS savings are not jeopardized by a minor equipment failure. The reputation that electronic equipment has earned over the years for good reliable service is well deserved. However that does not mean that it can be neglected. Hence a regular physical maintenance and inspection program should be developed along with an operating budget for repairs which may be necessary. This issue is discussed further in Chapter 14.

Security Issues

This discussion of security issues, translating to missed cost savings, is confined to system inefficiencies which are introduced via "unauthorized" and "uneducated" interface with the equipment. Such problems are limited to those exchanges which occur at the local panel itself, and those which may occur via front end. With respect to local panel exchanges, "unauthorized" refers to any type of system exchange that occurs through unsanctioned channels.

It is commonly known that manufacturers generally provide for three "password" levels of interface or access to their systems: "Read Only," "Read and Program," "Read, Program and Change Password." Here, it is assumed that the individuals conducting the interchange do not have terminal or site interface passwords. This category includes the classic EMS savings deterrents such as, pulling the plug, fuse, etc., and could also include electrical recircuiting to bypass the EMS interface.

Often, the severity of this problem is determined by the amount of direct local presence the energy manager has on site. It is assumed that most users will deal with uneducated exchanges from time to time, during training of new employees, etc. These are a primary source of lost savings. These usually involve features which have been provided with the local system apparatus to allow site interface. Often these are overrides or bypasses which are abused.

It is also possible to have an uneducated interface with system software. This happens when site personnel are provided with a site terminal and a password, allowing them to make changes that affect system integrity. Further discussion related to this concept will be addressed under performance and programming style.

The topic of PC-compatible front ends and remote system communication introduces a number of interesting variables into the discussion of unauthorized exchanges. Normally, front-end systems are password-protected, thus limiting unauthorized interface. Unauthorized interface can be severely limited by careful maintenance of password integrity. Uneducated interface is often a greater problem from remote communication. There are several considerations under this heading which will be discussed under system performance, as it is also necessary to consider programming access and style issues. The primary point is that interchanges should be monitored, and that individuals responsible for this activity should be trained, supervised and follow an established procedure to minimize problems stemming from this cause.

Performance

System performance is a broad category, and is affected by many variables. Assume that the actions recommended thus far have been carried out, so that the discussion may be limited to software.

First, let's review software access and style. Consider that a segment of the equipment available on the market allows limited user interface. Many minicomputer- and microcomputer-based systems employ proprietary programming languages. In such cases manufacturers often limit the user interface to: "Read Only" data and "Program Parameter" passwords.

These will allow a user to access data and change setpoints, but not to modify control algorithms, etc. This chapter addresses several specific problem areas, including software performance, that are germane for systems which allow direct user access to all or a portion of the system programming. The concept of system programming, however, is not to be confused with the manufacturer's source code. Source code is the manufacturer's proprietary program instructions, and is often called firmware.

System software of interest here is the user-level programming language, and may be viewed in two general categories: flexible and structured. As a clarification, all languages are structured, but the EMS firmware provides a language for user interface which will vary in format and style. Structured styles provide very specific formats which require a response to a prompt. Such systems are usually menu-driven and employ error trapping which limits responses to a particular prompt to answers within a specific range.

Flexible programming styles vary in degrees. For example, some systems will provide a library of strategies for the programmer to draw upon or link together, and incorporate into an individual control program. Others require programs or equations to accomplish each specific task. Each of the categories outlined under performance applies to both flexible and structured program style, but the extent of the effect will vary.

As with global system issues, software performance factors will be examined in two general categories: Verification and Fine-tuning. Verification is a check on software integrity which involves system update maintenance, and software problems through programming errors and software inconsistencies. Fine-tuning is a process of improving efficiency through adaptive programming modifications based on the examination of historical data.

System update integrity refers to changes in parameters that are stored permanently on disk, and noted to all system users. This should be a part of general software maintenance, yet it becomes complicated when more than one front end is employed. This is even more difficult when geographic distance is applied, and individuals who make changes to system parameters and strategies are not diligent about communicating those changes to other users.

A further complication occurs if for some reason the program residing in the EMS is lost, due to extended power failure, etc., and a new program must be downloaded. This situation can result in the need to retrace every step taken in the process of fine tuning and modifying the original program, a time-intensive process. Equally important to the energy manager is the fact that such occurrences often erode the facility user's faith in the EMS system.

Program integrity is a problem which is caused by software inconsistencies of several types. Basically, this problem occurs when there are errors made in programming. The initial program is modified incorrectly or a parameter input to the program is inconsistent with related control strategies. These concerns can be acute where flexible program styles are employed, and if there are no internal error trapping routines to guide the programmer.

Often such problems are the result of programming that is not proof-read adequately. However, depending upon the data checking features employed, this can also happen when programs are transmitted via modem. Interference or noise introduced to the line during transmission can result in a minor transposition of characters which is not picked up by the front end, and renders the control strategy ineffective.

Management Reporting

Reporting is an essential function to program management. It is necessary to produce reports on the fast track while maintaining a high level of quality. There are two general categories of reports: Operating Reports and Management Reports. Operating reports are normally of significance to the energy management staff only. These typically provide hard data for use in making day-to-day decisions, evaluating program performance, and justifying new projects.

Management reports summarily review major milestones and provide bottom line results. These reports will often employ charts, tables and other graphic displays of the information. Their primary function is to present program results to upper management in a concise manner.

SYSTEM FINE TUNING

System fine tuning differs from the activities discussed thus far in that it addresses modifications to software. The system verification issues discussed above are oriented towards maintaining the EMS in an optimum working environment, and ensuring integrity of the original programming. Fine tuning, as the term implies, is a process of evaluating the existing program instruction to determine whether control may be implemented more efficiently. This requires that specific processes be pinpointed for examination, and the following steps are carried out:

1.   Interrogate the system to determine current operating characteristics,

2.   Trend logs of historical data are developed and monitored,

3.   Opportunities for control enhancements are targeted,

4. Make calculations or assumptions based upon the above and modify program instructions,

5. Implement on a test basis in a system without controlled loads,

6. Field implementation,

7. Evaluate results.

This process should be conducted periodically on all controlled loads.

PROGRAM MANAGEMENT FUNCTIONS

In the previous sections, tasks were discussed rather than the functions or tools necessary to complete the tasks. There are in fact tools for accomplishing these tasks, and we will briefly discuss these in the remainder of this chapter. There are four primary functions, and it should be noted that currently there is not a fully-automated software or hardware system which incorporates all of them. As a result these functions are carried out through a hybrid combination of manual and automated steps. Chapter 11 on Communications and Standardization gives more information on this subject.

The basic program management functions to be completed are: Communications, Database Management, Analysis, and Reporting. Each of these functions has been discussed in varying detail throughout this book.

Communications is defined as the process of human interface with the EMS firmware. It is the elementary requirement to execute both system verification and performance tasks, because interface with the EMS is basic to program management.

Database management is the process whereby data is prepared for management review. This may entail sorting data by category, etc., placing it in arrays, summarizing and preparing data for analysis. It is essential to the performance aspects of program management.

Analysis is the process of conducting engineering or other operations on the above information. This function is the primary building block for system performance enhancements and in particular, fine tuning. The byproduct of these processes is to provide management with "Decision Data."

Reporting takes decision data and presents it for both in-house use and for management summaries of program status. Though this function is not directly utilized in program management, it is the cornerstone of a successful EMCS program.

Successful implementation of an EMCS program requires an effective management effort to address each of the issues presented in this chapter. As discussed in Chapter 2, lack of program management is one of the key reasons that some EMC systems do not meet their performance objective. Therefore, it is essential to establish such a program, and verify its effectiveness, on a continuous basis, with dedicated personnel.

18

EMC System Guidelines for New Buildings

EMC systems should be specified for all new projects which have:

1. More than 8,000 square feet of conditioned (heated and/or cooled) space,

2. more than 10,000 square feet of space which is heated only,

3. 20 or more tons of cooling with compressors of 5-ton capacity (or more) each,

4. heating capacity in excess of 500,000 Btu/hour,

5. a need for sophisticated control of HVAC, cogeneration, or other building systems,

6. a need for remote communications to ensure building performance and efficiency, or

7. a need for enhanced control, information management or system integration functions.

These criteria should not be taken to exclude those facilities which utilize a substantial amount of energy and for which application of EMC systems can save energy or money.

Applying EMC devices or systems to a new building involves many of the same concerns that affect existing buildings. The difference is, of course, that building design factors themselves can be adjusted to accommodate system requirements. This provides a high level of flexibility that can:

1. Increase building/EMC compatibility,

2. increase EMC effectiveness and optimize efficiency, and

3. lower equipment and installation cost.

The only advantage of working with an existing building is that various energy consumption and related data are available for purposes of analysis. This is not that much of an advantage, however, because virtually all aspects of building performance can be simulated, based on building design or alternative designs. In fact, use of building performance simulation is highly recommended during building design to help ensure that the most efficient systems are specified.

Some basic design guidelines for energy conservation are indicated below. If an EMC system or device is to realize maximum benefits, the building itself should be designed for maximum energy efficiency to begin with. In that regard, the controls provided are of extreme importance.

1. Study local utility rate structures, demand and power factor related requirements, likely future changes, etc.

2. Identify energy consuming devices that can be controlled.

3. Identify the most energy efficient control scheme, consistent with occupancy, weather, usage, etc., for each system and piece of equipment.

4. Design cost-effective systems that use the least amount of energy while being consistent with building needs. Energy and cost-effectiveness of various systems can be studied through computer simulation. Examples of effective control schemes include, but are not limited to:

   a. Reliance on separate systems for noncritical areas that can be shut down during unoccupied hours, and for critical areas that require special environmental conditions.

   b. Providing controls to minimize reheat and simultaneous heating and cooling.

   c. Avoiding use of 100% or fixed outside air systems where possible.

   d. Providing modulating outside air/return air dampers in air handling system in order to incorporate economizer or enthalpy controls.

e. Using variable air volume systems.

f. Providing chilled water temperature reset controls, to allow chilled water temperature reset consistent with cooling demands. Reset controls may be based on outdoor air dry bulb, wet bulb or enthalpy.

k. A double bundle condenser system should be evaluated for use in high-rise and mid-rise buildings to recover waste-heat for use elsewhere.

l. Provide for submetering of major loads.

m. Detail future expansion phasing, if any, and define future control and systems interfacing.

When a Level III EMC system is involved, it is suggested that drawings and specifications for the system, as well as bidding procedures, be treated as separate entities, since those responsible for general mechanical and electrical systems installation usually are not also adept in the field of EMC systems. If they are, however, then they should be allowed to bid on the separate EMC package. In any event, it is essential that there be close coordination of all trades on the project, to help ensure that the EMC system is integrated into new building in as effective a manner as possible, with an eye toward accessibility to enable prompt and efficient maintenance.

In the event that the EMCS for some reason will not be installed at the time the building is built, it is essential to provide for certain basic EMC system concerns to facilitate subsequent installation. In such cases, the A/E should indicate on plans and provide specifications for:

1. Field Interface Devices which are located in an environmentally controlled area (temperature 45F–110F; relative humidity 0–95% nonsweating). FIDs should be accessible by maintenance personnel without disrupting the mission of the facility or area in which installed.

2. A conduit termination raceway which will receive all data collection terminal cabinet (DTC) and, if required, telephone backboard conduits. The raceway should be located adjacent

to FID and should allow convenient cable routing into the FID terminal boards.

3. A Data Collection Terminal Cabinet (DTC) similar to the Automatic Temperature Control (ATC) system cabinet for each ATC control panel. The DTC should have sufficient terminal block space for 150% of EMC system inputs and outputs specified.

4. A ½-inch conduit, with pull wire, from each DTC to the conduit termination raceway adjacent to the FID location.

5. A ½-inch conduit, with pull wire, from the building telephone backboard, if required, to the conduit termination raceway adjacent to the FID location.

6. For each EMC system input or output connection required, the A&E should include in the plans and specifications provisions for the ATC manufacturer to provide the necessary transducers to transmit to or receive from the terminal block of the DTC analog or binary signals with the following characteristics:

    a. Direct current with voltage within the range from 0 to 5 VDC, to 0 to 20 VDC.

    b. 4 to 20 milliampere current loop with 250 ohms nominal impedance, 500 ohms maximum.

    c. Snap acting, gold-plated wiping contact rated for 24 VDC.

The ATC manufacturer should wire to the DTC terminal block all indicating, alarm, and control functions indicated by A&E. Terminals in DTC should be permanently identified as to function, type and range of signal. The DTC terminal block is intended to serve as the division line between ATC manufacturer responsibility and EMC system manufacturer, with an ATC manufacturer or service contractor, if any, maintaining the local controls up to and including the DTC and terminal block. The EMC system service contractor, if any, will maintain from the DTC terminal block on into the EMC system.

19

Life Cycle Costing

Albert Thumann, *P.E., C.E.M.*

One of the most important steps in planning an EMC System is determining the economic basis for its purchase.

Some companies use a simple payback method of two years or less to justify equipment purchases. Others require a life cycle cost analysis with no fuel price inflation considered. Still other companies allow for a complete life cycle cost analysis, including the impact for the fuel price inflation.

The energy manager's success is directly related to how he or she must justify energy utilization methods.

USING THE PAYBACK PERIOD METHOD

The payback period is the time required to recover the capital investment out of the earnings or savings. This method ignores all savings beyond the payback years, thus penalizing projects that have long life potentials for those that offer high savings for a relatively short period.

The payback period criterion is used when funds are limited and it is important to know how fast dollars will come back. The payback period is simply computed as:

$$\text{Payback period} = \frac{\text{initial investment}}{\text{savings}} \qquad (19\text{-}1)$$

The energy manager who must justify energy equipment expenditures based on a payback period of one year or less has little chance for long-range success. Some companies have set higher payback periods for energy utilization methods. These longer payback periods are justified on the basis that:

- Fuel pricing will increase at a higher rate than the general inflation rate.

- The "risk analysis" for not implementing energy utilization measures may mean loss of production and losing a competitive edge.

USING LIFE CYCLE COSTING

Life cycle costing is an analysis of the total cost of a system, device, building, machine, etc., over its anticipated useful life. The subject has, in the past, gone by such names as "engineering economic analysis" or "total owning and operating cost summaries."

Life cycle costing has brought about a new emphasis on the comprehensive identification of all costs associated with a system. The most commonly included costs are initial in-place cost, operating costs, maintenance costs, and interest on the investment. Two factors enter into appraising the life of the system; namely, the expected physical life and the period of obsolescence. The lesser factor is governing time period. The effect of interest can then be calculated by using one of several formulas which take into account the time value of money.

When comparing alternative solutions to a particular problem the system showing the lowest life cycle cost will usually be the first choice (performance requirements are assessed as equal in value).

Life cycle costing is a tool in value engineering. Other items, such as installation time, pollution effects, aesthetic considerations, delivery time, and owner preferences will temper the rule of always choosing the system with the lowest life cycle cost. Good overall judgment is still required.

The life cycle cost analysis still contains judgment factors pertaining to interest rates, useful life, and inflation rates. Even with the judgment element, life cycle costing is the most important tool in value engineering, since the results are quantified in terms of dollars.

As the price for energy changes, and as governmental incentives are initiated, processes or alternatives which were not economically feasible will be considered. This chapter will concentrate on the principles of the life cycle cost analysis as they apply to energy conservation decision making.

THE TIME VALUE OF MONEY

Most energy saving proposals require the investment of capital to accomplish them. By investing today in energy conservation, yearly operating dollars over the life of the investment will be saved. A dollar in hand today is more valuable than one to be received at some time in the future. For this reason, a *time value* must be placed on all cash flows into and out of the company.

Money transactions are thought of as a cash flow to or from a company. Investment decisions also take into account alternate investment opportunities and the minimum return on the investment. In order to compute the rate of return on an investment, it is necessary to find the interest rate which equates payments outgoing and incoming, present and future. The method used to find the rate of return is referred to as *discounted cash flow.*

INVESTMENT DECISION-MAKING

To make investment decisions, the energy manager must follow one simple principle: Relate annual cash flows and lump sum deposits to the same time base. The six factors used for investment decision-making simply convert cash from one time base to another; since each company has various financial objectives, these factors can be used to solve *any* investment problem.

Single Payment Compound Amount—SPCA

The SPCA factor is used to determine the future amount S that a present sum P will accumulate at i percent interest, in n years. If P (present worth) is known, and S (future worth) is to be determined, then Equation 19-2 is used.

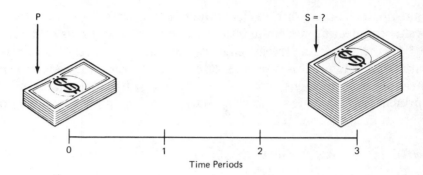

Time Periods

Fig. 19-1. Single Payment Compound Amount (SPCA)

$$S = P \times (SPCA)n_i \qquad (19\text{-}2)$$

$$SPCA = (1 + i)^n \qquad (19\text{-}3)$$

The SPCA can be computed by an interest formula, but usually its value is found by using the interest tables. Interest tables for interest rates of 10 to 50 percent are found at the conclusion of this chapter (Tables 19-2 through 19-9). In predicting future costs, there are many unknowns. For the accuracy of most calculations, interest rates are assumed to be compounded annually unless otherwise specified. Linear interpolation is commonly used to find values not listed in the interest tables.

Tables 19-10 through 19-13 can be used to determine the effect of fuel escalation on the life cycle cost analysis.

Single Payment Present Worth—SPPW

The SPPW factor is used to determine the present worth, P, that a future amount, S, will be at interest of i-percent, in n years. If S is known, and P is to be determined, then Equation 19-4 is used.

$$P = S \times (SPPW)i_n \qquad (19\text{-}4)$$

$$SPPW = \frac{1}{(1 + i)^n} \qquad (19\text{-}5)$$

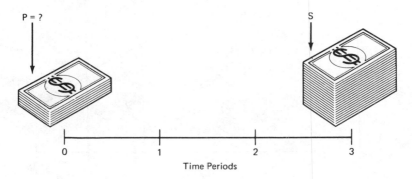

Fig. 19-2. Single Payment Present Worth (SPPW)

Uniform Series Compound Amount—USCA

The USCA factor is used to determine the amount S that an equal annual payment R will accumulate to in n years at i percent interest. If R (uniform annual payment) is known, and S (the future worth of these payments) is required, then Equation 19-6 is used.

$$S = R \times (USCA)i_n \qquad (19\text{-}6)$$

$$USCA = \frac{(1 + i)^n - 1}{i} \qquad (19\text{-}7)$$

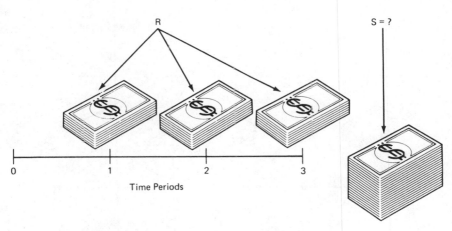

Fig. 19-3. Uniform Series Compound Amount (USCA)

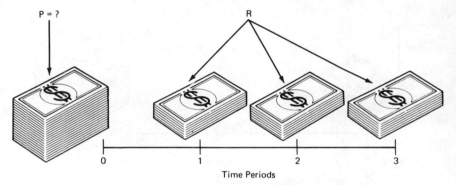

Fig. 19-4. Uniform Series Present Worth (USPW)

Uniform Series Present Worth–(USPW)

The USPW factor is used to determine the present amount P that can be paid by equal payments of R (uniform annual payment) at i percent interest, for n years. If R is known, and P is required, then Equation 19-8 is used.

$$P = R \times (USPW)i_n \qquad (19\text{-}8)$$

$$USPW = \frac{(1 + i)^n - 1}{i(1 + i)^n} \qquad (19\text{-}9)$$

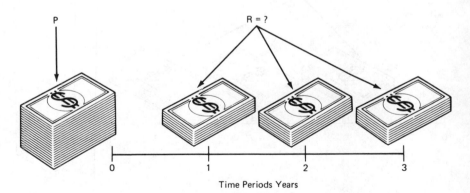

Fig. 19-5. Capital Recovery (CR)

Capital Recovery—CR

The CR factor is used to determine an annual payment R required to pay off a present amount P at i percent interest, for n years. If the present sum of money, P, spent today is known, and the uniform payment R needed to pay back P over a stated period of time is required, then Equation 19-10 is used.

$$R = P \times (CR)i_n \tag{19-10}$$

$$CR = \frac{i(1 + i)^n}{(1 + i)^n - 1} \tag{19-11}$$

Sinking Fund Payment—SFP

The SFP factor is used to determine the equal annual amount R that must be invested for n years at i percent interest in order to accumulate a specified future amount. If S (the future worth of a series of annual payments) is known, and R (value of those annual payments) is required, then Equation 19-12 is used.

$$R = S \times (SFP)i_n \tag{19-12}$$

$$SFP = \frac{i}{(1 + i)^n - 1} \tag{19-13}$$

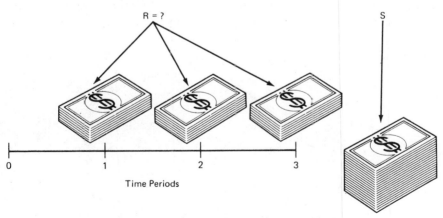

Fig. 19-6. Sinking Fund Payment (SFP)

Gradient Present Worth—GPW

The GPW factor is used to determine the present amount P that can be paid by annual amounts R which escalate at e percent, at i percent interest, for n years. If R is known, and P is required, then Equation 19-14 is used. The GPW factor is a relatively new term which has gained in importance due to the impact of inflation.

$$P = R \times (GPW)i_n \qquad (19\text{-}14)$$

$$GPW = \frac{\dfrac{1+e}{1+i}\left[1 - \left(\dfrac{1+e}{1+i}\right)^n\right]}{1 - \dfrac{1+e}{1+i}} \qquad (19\text{-}15)$$

The three most commonly used methods in life cycle costing are the annual cost, present worth and rate-of-return analysis.

In the present worth method a minimum rate of return (i) is stipulated. All future expenditures are converted to present values using the interest factors. The alternative with lowest effective first cost is the most desirable.

A similar procedure is implemented in the annual cost method. The difference is that the first cost is converted to an annual expenditure. The alternative with lowest effective annual cost is the most desirable.

In the rate-of-return method, a trial-and-error procedure is usually required. Interpolation from the interest tables can determine what rate of return (i) will give an interest factor which will make the

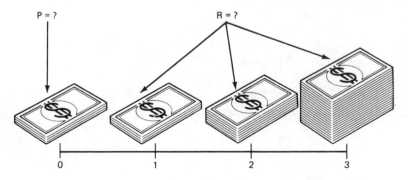

Fig. 19-7. Gradient Present Worth (GPW)

overall cash flow balance. The rate-of-return analysis gives a good indication of the overall ranking of independent alternates.

The effect of escalation in fuel costs can influence greatly the final decision. When an annual cost grows at a steady rate it may be treated as a gradient and the gradient present worth factor can be used.

Special thanks are given to Rudolph R. Yanuck and Dr. Robert Brown for the use of their specially designed interest and escalation tables used in this text.

When life cycle costing is used to compare several alternatives the differences between costs are important. For example, if one alternate forces additional maintenance or an operating expense to occur, then these factors as well as energy costs need to be included. Remember, what was previously spent for the item to be replaced is irrelevant. The only factor to be considered is whether the new cost can be justified based on projected savings over its useful life.

MAKING DECISIONS FOR ALTERNATE INVESTMENTS

There are several methods for determining which energy conservation alternative is the most economical. Probably the most familiar and trusted method is the annual cost method.

When evaluating replacement of processes or equipment *do not* consider what was previously spent. The decision will be based on whether the new process or equipment proves to save substantially enough in operating costs to justify the expenditure.

Equation 19-16 is used to convert the lump sum investment P into the annual cost. In the case where the asset has a value after the end of its useful life, the annual cost becomes

$$AC = (P - L) \, CR + iL \qquad (19\text{-}16)$$

where

AC is the annual cost

L is the net sum of money that can be realized for a piece of equipment, over and above its removal cost, when it is returned at the end of the service life. L is referred to as the salvage value.

As a practical point, the salvage value is usually small and can be neglected, considering the accuracy of future costs. The annual cost technique can be implemented by using the following format:

	Alternate 1	Alternate 2
1. First cost *(P)*		
2. Estimated life *(n)*		
3. Estimated salvage value at end of life *(L)*		
4. Annual disbursements, including energy costs & maintenance *(E)*		
5. Minimum acceptable return *before* taxes *(i)*		
6. CR *n, i*		
7. *(P − L)* CR		
8. *Li*		
9. AC = *(P − L)* CR + *Li* + *E*		

Choose alternate with lowest AC

The alternative with the lowest annual cost is the desired choice.

DEPRECIATION, TAXES, AND THE TAX CREDIT

Depreciation

Depreciation affects the "accounting procedure" for determining profits and losses and the income tax of a company. In other words, for tax purposes the expenditure for an asset such as a pump or motor cannot be fully expensed in its first year. The original investment must be charged off for tax purposes over the useful life of the asset. A company usually wishes to expense an item as quickly as possible.

The Internal Revenue Service allows several methods for determining the annual depreciation rate.

Straight-Line Depreciation. The simplest method is referred to as a straight-line depreciation and is defined as:

$$D = \frac{P - L}{n}$$ (19-17)

where

D is the annual depreciation rate
L is the value of equipment at the end of its useful life, common-
   ly referred to as salvage value
n is the life of the equipment, which is determined by Internal
   Revenue Service guidelines
P is the initial expenditure.

Sum-of-Years Digits. Another method is referred to as the sum-of-
years digits. In this method the depreciation rate is determined by
finding the sum of digits using the following formula,

$$N = n\frac{(n + 1)}{2}$$ (19-18)

where *n* is the life of equipment.
   Each year's depreciation rate is determined as follows:

First year $$D = \frac{n}{N}(P - L)$$ (19-19)

Second year $$D = \frac{n - 1}{N}(P - L)$$ (19-20)

n year $$D = \frac{1}{N}(P - L)$$ (19-21)

Declining-Balance Depreciation. The declining-balance method
allows for larger depreciation charges in the early years which is
sometimes referred to as fast write-off.
   The rate is calculated by taking a constant percentage of the declining undepreciated balance. The most common method used to
calculate the declining balance is to predetermine the depreciation
rate. Under certain circumstances a rate equal to 200 percent of the
straight-line depreciation rate may be used. Under other circumstances the rate is limited to 1½ or ¼ times as great as straight-line
depreciation. In this method the salvage value or undepreciated book
value is established once the depreciation rate is preestablished.

To calculate the undepreciated book value, Equation 19-22 is used.

$$D = 1 - \left(\frac{L}{P}\right)^{1/N} \tag{19-22}$$

where

D is the annual depreciation rate
L is the salvage value
P is the first cost.

As a result of The Economic Recovery Tax Act of 1981 there is a generally faster method of writing off the cost of tangible property used in business or held for the production of income. It's called the "Accelerated Cost Recovery System" or ACRS. The new system is generally applicable to eligible property (called "recovery property") placed in service on or after January 1, 1981. So it may apply to depreciable property that you've already purchased.

Classes of Recovery Property

Recovery property is divided into four classes: 3-year, 5-year, 10-year, and 15-year property. For example:

- 3-year: Cars, light duty trucks and certain other short-lived personal property.
- 5-year: Most machinery and equipment.
- 15-year: Buildings.

For each class there is a standard set of recovery deductions (i.e., depreciation with a new name) to be taken over a fixed recovery period.

Tax Considerations

The Income Tax Reform Act of 1986 has impacted many of the decisions that building owners make with regard to capital modification. It is recommended that your corporate accountant be consulted to address the tax considerations associated with that investment under the new law. As a rule, expenses such as maintenance, energy, operating costs, insurance, and property taxes, are tax deductible, and reduce the income subject to taxes.

For the after-tax life cycle cost analysis and payback analysis the actual incurred and annual savings is given as follows:

$$AS = (1 - I)E + ID \qquad (19\text{-}23)$$

where

AS is the yearly annual after tax savings (excluding effect of tax credit)

E is the yearly annual energy savings (difference between original expenses and expenses after modification)

D is the annual depreciation rate

I is the income tax bracket.

Equation 19-23 takes into account that the yearly annual energy savings is partially offset by additional taxes which must be paid due to reduced operating expenses. On the other hand, the depreciation allowance reduces taxes directly.

Tax Credit

Today tax credits are no longer a reality with regard to the typical EMCS investment. They will be addressed in this section for information purposes, and to document the program which was formerly in place.

A tax credit encourages capital investment. Essentially the tax credit lowers the income tax paid by the tax credit to an upper limit. Two basic credits were available: investment tax credit, and the Business Energy Tax Credit which applied to industrial investment in alternative energy property such as boilers for coal, heat conservation, and recycling equipment. These tax credits substantially increased the investment merit of the investment since it lowered the *bottom* line on the tax form. As noted, EMC systems implemented today must be justified on the merits of energy savings and non-energy related benefits such as information management and system integration.

After-Tax Analysis

To compute a rate of return which accounts for taxes, depreciation, escalation, and tax credits, a cash-flow analysis is usually required. This method analyzes all transactions including first and operating costs. To determine the after-tax rate of return a trial and error or computer analysis is required.

All money is converted to the present assuming an interest rate. The summation of all present dollars should equal zero when the correct interest rate is selected, as illustrated in Figure 19-8.

This analysis can be made assuming a fuel escalation rate by using the gradient present worth interest of the present worth factor.

Figure 19-9 illustrates the effects of escalation. This figure can be used as a quick way to determine after-tax economics of energy utilization expenditures.

	1	2	3	4	(2 + 3) × 4
Year	Investment	Tax Credit	After Tax Savings (AS)	Single Payment Present Worth Factor	Present Worth
0	$-P$				$-P$
1		$+TC$	AS	$SPPW_1$	$+P_1$
2			AS	$SPPW_2$	P_2
3			AS	$SPPW_3$	P_3
4			AS	$SPPW_4$	P_4
Total					ΣP

$$AS = (1 - I)E + ID$$
Trial and Error Solution:
Correct i when $\Sigma P = 0$

Fig. 19-8. Cash Flow Rate of Return Analysis

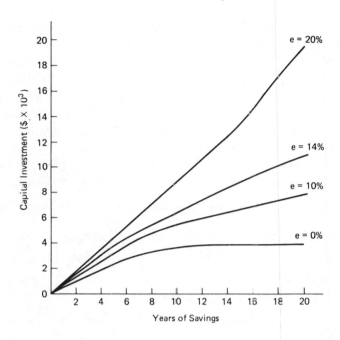

Fig. 19-9. Effects of Escalation

LIFE CYCLE COST ANALYSIS OPTIMIZATION

Plant modifications and expansions have been traditionally governed by an emphasis on first cost. Today's decisions are governed by the life cycle cost analysis approach. This analysis forces predictions as to:

1. Total energy consumed
2. Cost of energy
3. Operation and maintenance procedures
4. Equipment replacement

The optimum savings opportunity will be found by looking at various alternatives. Any life cycle cost analysis can be optimized by finding the lowest annual cost which is bounded by two higher costs.

The principles of life cycle costing should be used to "sell" energy utilization proposals within an organization. Energy utilization proposals should stress not what the first cost is, but rather how much will be saved each year and the return on the investment.

Another interesting relationship is the sensitivity of variables, such as interest rates, on the life cycle cost analysis. The checklist of Table 19-1 illustrates the effect of variables on cost decisions. To measure the impact of these variables, the sensitivity of the end result should be tested by varying a questionable element (upward or downward) and redoing the calculation.

Table 19-1. Life Cycle Cost-Variable Sensitivity Checklist

1. Low interest rates de-emphasize a high initial cost.
2. High interest rates favor a system with low first cost.
3. Short life favors low first cost.
4. Long life de-emphasizes high initial cost.
5. Short annual operating hours de-emphasize high energy rates or low system efficiency.
6. Long annual operating hours favor a system with high efficiency.
7. Low average load factor de-emphasizes low system efficiency.
8. High average load factor favors a system with high efficiency.

Life cycle costing is a new name for a basic engineering economics technique. Before it is applied, it needs to be understood by all concerned.

Life cycle costing should be used in energy managers' recommendations and should become a part of the engineering analysis.

Table 19-2. 10% Interest Factors

Period n	Single-payment compound-amount (SPCA)	Single-payment present-worth (SPPW)	Uniform-series compound-amount (USCA)	Sinking-fund payment (SFP)	Capital recovery (CR)	Uniform-series present-worth (USPW)
	Future value of $1 $(1+i)^n$	Present value of $1 $\frac{1}{(1+i)^n}$	Future value of uniform series of $1 $\frac{(1+i)^n-1}{i}$	Uniform series whose future value is $1 $\frac{i}{(1+i)^n-1}$	Uniform series with present value of $1 $\frac{i(1+i)^n}{(1+i)^n-1}$	Present value of uniform series of $1 $\frac{(1+i)^n-1}{i(1+i)^n}$
1	1.100	0.9091	1.000	1.00000	1.10000	0.909
2	1.210	0.8264	2.100	0.47619	0.57619	1.736
3	1.331	0.7513	3.310	0.30211	0.40211	2.487
4	1.464	0.6830	4.641	0.21547	0.31547	3.170
5	1.611	0.6209	6.105	0.16380	0.26380	3.791
6	1.772	0.5645	7.713	0.12961	0.22961	4.355
7	1.949	0.5132	9.487	0.10541	0.20541	4.868
8	2.144	0.4665	11.436	0.08744	0.18744	5.335
9	2.358	0.4241	13.579	0.07364	0.17364	5.759
10	2.594	0.3855	15.937	0.06275	0.16275	6.144
11	2.853	0.3505	18.531	0.05396	0.15396	6.495
12	3.138	0.3186	21.384	0.04676	0.14676	6.814
13	3.452	0.2897	24.523	0.04078	0.14078	7.103
14	3.797	0.2633	27.975	0.03575	0.13575	7.367
15	4.177	0.2394	31.772	0.03147	0.13147	7.606
16	4.595	0.2176	35.950	0.02782	0.12782	7.824
17	5.054	0.1978	40.545	0.02466	0.12466	8.022
18	5.560	0.1799	45.599	0.02193	0.12193	8.201
19	6.116	0.1635	51.159	0.01955	0.11955	8.365
20	6.727	0.1486	57.275	0.01746	0.11746	8.514
21	7.400	0.1351	64.002	0.01562	0.11562	8.649
22	8.140	0.1228	71.403	0.01401	0.11401	8.772
23	8.954	0.1117	79.543	0.01257	0.11257	8.883
24	9.850	0.1015	88.497	0.01130	0.11130	8.985
25	10.835	0.0923	98.347	0.01017	0.11017	9.077
26	11.918	0.0839	109.182	0.00916	0.10916	9.161
27	13.110	0.0763	121.100	0.00826	0.10826	9.237
28	14.421	0.0693	134.210	0.00745	0.10745	9.307
29	15.863	0.0630	148.631	0.00673	0.10673	9.370
30	17.449	0.0573	164.494	0.00608	0.10608	9.427
35	28.102	0.0356	271.024	0.00369	0.10369	9.644
40	45.259	0.0221	442.593	0.00226	0.10226	9.779
45	72.890	0.0137	718.905	0.00139	0.10139	9.863
50	117.391	0.0085	1163.909	0.00086	0.10086	9.915
55	189.059	0.0053	1880.591	0.00053	0.10053	9.947
60	304.482	0.0033	3034.816	0.00033	0.10033	9.967
65	490.371	0.0020	4893.707	0.00020	0.10020	9.980
70	789.747	0.0013	7887.470	0.00013	0.10013	9.987
75	1271.895	0.0008	12708.954	0.00008	0.10008	9.992
80	2048.400	0.0005	20474.002	0.00005	0.10005	9.995
85	3298.969	0.0003	32979.690	0.00003	0.10003	9.997
90	5313.023	0.0002	53120.226	0.00002	0.10002	9.998
95	8556.676	0.0001	85556.760	0.00001	0.10001	9.999

Table 19-3. 12% Interest Factors

Period n	Single-payment compound-amount (SPCA) Future value of $1 $(1+i)^n$	Single-payment present-worth (SPPW) Present value of $1 $\dfrac{1}{(1+i)^n}$	Uniform-series compound-amount (USCA) Future value of uniform series of $1 $\dfrac{(1+i)^n-1}{i}$	Sinking-fund payment (SFP) Uniform series whose future value is $1 $\dfrac{i}{(1+i)^n-1}$	Capital recovery (CR) Uniform series with present value of $1 $\dfrac{i(1+i)^n}{(1+i)^n-1}$	Uniform-series present-worth (USPW) Present value of uniform series of $1 $\dfrac{(1+i)^n-1}{i(1+i)^n}$
1	1.120	0.8929	1.000	1.00000	1.12000	0.893
2	1.254	0.7972	2.120	0.47170	0.59170	1.690
3	1.405	0.7118	3.374	0.29635	0.41635	2.402
4	1.574	0.6355	4.779	0.20923	0.32923	3.037
5	1.762	0.5674	6.353	0.15741	0.27741	3.605
6	1.974	0.5066	8.115	0.12323	0.24323	4.111
7	2.211	0.4523	10.089	0.09912	0.21912	4.564
8	2.476	0.4039	12.300	0.08130	0.20130	4.968
9	2.773	0.3606	14.776	0.06768	0.18768	5.328
10	3.106	0.3220	17.549	0.05698	0.17698	5.650
11	3.479	0.2875	20.655	0.04842	0.16842	5.938
12	3.896	0.2567	24.133	0.04144	0.16144	6.194
13	4.363	0.2292	28.029	0.03568	0.15568	6.424
14	4.887	0.2046	32.393	0.03087	0.15087	6.628
15	5.474	0.1827	37.280	0.02682	0.14682	6.811
16	6.130	0.1631	42.753	0.02339	0.14339	6.974
17	6.866	0.1456	48.884	0.02046	0.14046	7.120
18	7.690	0.1300	55.750	0.01794	0.13794	7.250
19	8.613	0.1161	63.440	0.01576	0.13576	7.366
20	9.646	0.1037	72.052	0.01388	0.13388	7.469
21	10.804	0.0926	81.699	0.01224	0.13224	7.562
22	12.100	0.0826	92.503	0.01081	0.13081	7.645
23	13.552	0.0738	104.603	0.00956	0.12956	7.718
24	15.179	0.0659	118.155	0.00846	0.12846	7.784
25	17.000	0.0588	133.334	0.00750	0.12750	7.843
26	19.040	0.0525	150.334	0.00665	0.12665	7.896
27	21.325	0.0469	169.374	0.00590	0.12590	7.943
28	23.884	0.0419	190.699	0.00524	0.12524	7.984
29	26.750	0.0374	214.583	0.00466	0.12466	8.022
30	29.960	0.0334	241.333	0.00414	0.12414	8.055
35	52.800	0.0189	431.663	0.00232	0.12232	8.176
40	93.051	0.0107	767.091	0.00130	0.12130	8.244
45	163.988	0.0061	1358.230	0.00074	0.12074	8.283
50	289.002	0.0035	2400.018	0.00042	0.12042	8.304
55	509.321	0.0020	4236.005	0.00024	0.12024	8.317
60	897.597	0.0011	7471.641	0.00013	0.12013	8.324
65	1581.872	0.0006	13173.937	0.00008	0.12008	8.328
70	2787.800	0.0004	23223.332	0.00004	0.12004	8.330
75	4913.056	0.0002	40933.799	0.00002	0.12002	8.332
80	8658.483	0.0001	72145.692	0.00001	0.12001	8.332

Table 19-4. 15% Interest Factors

Period *n*	Single-payment compound-amount (SPCA) Future value of $1 $(1 + i)^n$	Single-payment present-worth (SPPW) Present value of $1 $\dfrac{1}{(1 + i)^n}$	Uniform-series compound-amount (USCA) Future value of uniform series of $1 $\dfrac{(1 + i)^n - 1}{i}$	Sinking-fund payment (SFP) Uniform series whose future value is $1 $\dfrac{i}{(1 + i)^n - 1}$	Capital recovery (CR) Uniform series with present value of $1 $\dfrac{i(1 + i)^n}{(1 + i)^n - 1}$	Uniform-series present-worth (USPW) Present value of uniform series of $1 $\dfrac{(1 + i)^n - 1}{i(1 + i)^n}$
1	1.150	0.8696	1.000	1.00000	1.15000	0.870
2	1.322	0.7561	2.150	0.46512	0.61512	1.626
3	1.521	0.6575	3.472	0.28798	0.43798	2.283
4	1.749	0.5718	4.993	0.20027	0.35027	2.855
5	2.011	0.4972	6.742	0.14832	0.29832	3.352
6	2 313	0.4323	8.754	0.11424	0.26424	3.784
7	2 660	0.3759	11.067	0.09036	0.24036	4.160
8	3 059	0.3269	13.727	0.07285	0.22285	4.487
9	3.518	0.2843	16.786	0.05957	0.20957	4.772
10	4.046	0.2472	20.304	0.04925	0.19925	5.019
11	4.652	0.2149	24.349	0.04107	0.19107	5.234
12	5.350	0.1869	29.002	0.03448	0.18448	5.421
13	6.153	0.1625	34.352	0.02911	0.17911	5.583
14	7.076	0.1413	40.505	0.02469	0.17469	5.724
15	8.137	0.1229	47.580	0.02102	0.17102	5.847
16	9.358	0.1069	55.717	0.01795	0.16795	5.954
17	10.761	0.0929	65.075	0.01537	0.16537	6.047
18	12.375	0.0808	75.836	0.01319	0.16319	6.128
19	14.232	0.0703	88.212	0.01134	0.16134	6.198
20	16.367	0.0611	102.444	0.00976	0.15976	6.259
21	18.822	0.0531	118.810	0.00842	0.15842	6.312
22	21.645	0.0462	137.632	0.00727	0.15727	6.359
23	24.891	0.0402	159.276	0.00628	0.15628	6.399
24	28.625	0.0349	184.168	0.00543	0.15543	6.434
25	32.919	0.0304	212.793	0.00470	0.15470	6.464
26	37.857	0.0264	245.712	0.00407	0.15407	6.491
27	43.535	0.0230	283.569	0.00353	0.15353	6.514
28	50.066	0.0200	327.104	0.00306	0.15306	6.534
29	57.575	0.0174	377.170	0.00265	0.15265	6.551
30	66.212	0.0151	434.745	0.00230	0.15230	6.566
35	133.176	0.0075	881.170	0.00113	0.15113	6.617
40	267.864	0.0037	1779.090	0.00056	0.15056	6.642
45	538.769	0.0019	3585.128	0.00028	0.15028	6.654
50	1083.657	0.0009	7217.716	0.00014	0.15014	6.661
55	2179.622	0.0005	14524.148	0.00007	0.15007	6.664
60	4383.999	0.0002	29219.992	0.00003	0.15003	6.665
65	8817.787	0.0001	58778.583	0.00002	0.15002	6.666

Table 19-5. 20% Interest Factors

Period n	Single-payment compound-amount (SPCA)	Single-payment present-worth (SPPW)	Uniform-series compound-amount (USCA)	Sinking-fund payment (SFP)	Capital recovery (CR)	Uniform-series present-worth (USPW)
	Future value of $1 $(1 + i)^n$	Present value of $1 $\dfrac{1}{(1 + i)^n}$	Future value of uniform series of $1 $\dfrac{(1 + i)^n - 1}{i}$	Uniform series whose future value is $1 $\dfrac{i}{(1 + i)^n - 1}$	Uniform series with present value of $1 $\dfrac{i(1 + i)^n}{(1 + i)^n - 1}$	Present value of uniform series of $1 $\dfrac{(1 + i)^n - 1}{i(1 + i)^n}$
1	1.200	0.8333	1.000	1.00000	1.20000	0.833
2	1.440	0.6944	2.200	0.45455	0.65455	1.528
3	1.728	0.5787	3.640	0.27473	0.47473	2.106
4	2.074	0.4823	5.368	0.18629	0.38629	2.589
5	2.488	0.4019	7.442	0.13438	0.33438	2.991
6	2.986	0.3349	9.930	0.10071	0.30071	3.326
7	3.583	0.2791	12.916	0.07742	0.27742	3.605
8	4.300	0.2326	16.499	0.06061	0.26061	3.837
9	5.160	0.1938	20.799	0.04808	0.24808	4.031
10	6.192	0.1615	25.959	0.03852	0.23852	4.192
11	7.430	0.1346	32.150	0.03110	0.23110	4.327
12	8.916	0.1122	39.581	0.02526	0.22526	4.439
13	10.699	0.0935	48.497	0.02062	0.22062	4.533
14	12.839	0.0779	59.196	0.01689	0.21689	4.611
15	15.407	0.0649	72.035	0.01388	0.21388	4.675
16	18.488	0.0541	87.442	0.01144	0.21144	4.730
17	22.186	0.0451	105.931	0.00944	0.20944	4.775
18	26.623	0.0376	128.117	0.00781	0.20781	4.812
19	31.948	0.0313	154.740	0.00646	0.20646	4.843
20	38.338	0.0261	186.688	0.00536	0.20536	4.870
21	46.005	0.0217	225.026	0.00444	0.20444	4.891
22	55.206	0.0181	271.031	0.00369	0.20369	4.909
23	66.247	0.0151	326.237	0.00307	0.20307	4.925
24	79.497	0.0126	392.484	0.00255	0.20255	4.937
25	95.396	0.0105	471.981	0.00212	0.20212	4.948
26	114.475	0.0087	567.377	0.00176	0.20176	4.956
27	137.371	0.0073	681.853	0.00147	0.20147	4.964
28	164.845	0.0061	819.223	0.00122	0.20122	4.970
29	197.814	0.0051	984.068	0.00102	0.20102	4.975
30	237.376	0.0042	1181.882	0.00085	0.20085	4.979
35	590.668	0.0017	2948.341	0.00034	0.20034	4.992
40	1469.772	0.0007	7343.858	0.00014	0.20014	4.997
45	3657.262	0.0003	18281.310	0.00005	0.20005	4.999
50	9100.438	0.0001	45497.191	0.00002	0.20002	4.999

Table 19-6. 25% Interest Factors

Period n	Single-payment compound-amount (SPCA) Future value of $1 $(1 + i)^n$	Single-payment present-worth (SPPW) Present value of $1 $\dfrac{1}{(1 + i)^n}$	Uniform-series compound amount (USCA) Future value of uniform series of $1 $\dfrac{(1 + i)^n - 1}{i}$	Sinking-fund payment (SFP) Uniform series whose future value is $1 $\dfrac{i}{(1 + i)^n - 1}$	Capital recovery (CR) Uniform series with present value of $1 $\dfrac{i(1 + i)^n}{(1 + i)^n - 1}$	Uniform-series present-worth (USPW) Present value of uniform series of $1 $\dfrac{(1 + i)^n - 1}{i(1 + i)^n}$
1	1.250	0.8000	1.000	1.00000	1.25000	0.800
2	1.562	0.6400	2.250	0.44444	0.69444	1.440
3	1.953	0.5120	3.812	0.26230	0.51230	1.952
4	2.441	0.4096	5.766	0.17344	0.42344	2.362
5	3.052	0.3277	8.207	0.12185	0.37185	2.689
6	3.815	0.2621	11.259	0.08882	0.33882	2.951
7	4.768	0.2097	15.073	0.06634	0.31634	3.161
8	5.960	0.1678	19.842	0.05040	0.30040	3.329
9	7.451	0.1342	25.802	0.03876	0.28876	3.463
10	9.313	0.1074	33.253	0.03007	0.28007	3.571
11	11.642	0.0859	42.566	0.02349	0.27349	3.656
12	14.552	0.0687	54.208	0.01845	0.26845	3.725
13	18.190	0.0550	68.760	0.01454	0.26454	3.780
14	22.737	0.0440	86.949	0.01150	0.26150	3.824
15	28.422	0.0352	109.687	0.00912	0.25912	3.859
16	35.527	0.0281	138.109	0.00724	0.25724	3.887
17	44.409	0.0225	173.636	0.00576	0.25576	3.910
18	55.511	0.0180	218.045	0.00459	0.25459	3.928
19	69.389	0.0144	273.553	0.00366	0.25366	3.942
20	86.736	0.0115	342.945	0.00292	0.25292	3.954
21	108.420	0.0092	429.681	0.00233	0.25233	3.963
22	135.525	0.0074	538.101	0.00186	0.25186	3.970
23	169.407	0.0059	673.626	0.00148	0.25148	3.976
24	211.758	0.0047	843.033	0.00119	0.25119	3.981
25	264.698	0.0038	1054.791	0.00095	0.25095	3.985
26	330.872	0.0030	1319.489	0.00076	0.25076	3.988
27	413.590	0.0024	1650.361	0.00061	0.25061	3.990
28	516.988	0.0019	2063.952	0.00048	0.25048	3.992
29	646.235	0.0015	2580.939	0.00039	0.25039	3.994
30	807.794	0.0012	3227.174	0.00031	0.25031	3.995
35	2465.190	0.0004	9856.761	0.00010	0.25010	3.998
40	7523.164	0.0001	30088.655	0.00003	0.25003	3.999

Table 19-7. 30% Interest Factors

Period n	Single-payment compound-amount (SPCA) Future value of $1 $(1 + i)^n$	Single-payment present-worth (SPPW) Present value of $1 $\dfrac{1}{(1 + i)^n}$	Uniform-series compound-amount (USCA) Future value of uniform series of $1 $\dfrac{(1 + i)^n - 1}{i}$	Sinking-fund payment (SFP) Uniform series whose future value is $1 $\dfrac{i}{(1 + i)^n - 1}$	Capital recovery (CR) Uniform series with present value of $1 $\dfrac{i(1 + i)^n}{(1 + i)^n - 1}$	Uniform-series present-worth (USPW) Present value of uniform series of $1 $\dfrac{(1 + i)^n - 1}{i(1 + i)^n}$
1	1.300	0.7692	1.000	1.00000	1.30000	0.769
2	1.690	0.5917	2.300	0.43478	0.73478	1.361
3	2.197	0.4552	3.990	0.25063	0.55063	1.816
4	2.856	0.3501	6.187	0.16163	0.46163	2.166
5	3.713	0.2693	9.043	0.11058	0.41058	2.436
6	4.827	0.2072	12.756	0.07839	0.37839	2.643
7	6.275	0.1594	17.583	0.05687	0.35687	2.802
8	8.157	0.1226	23.858	0.04192	0.34192	2.925
9	10.604	0.0943	32.015	0.03124	0.33124	3.019
10	13.786	0.0725	42.619	0.02346	0.32346	3.092
11	17.922	0.0558	56.405	0.01773	0.31773	3.147
12	23.298	0.0429	74.327	0.01345	0.31345	3.190
13	30.288	0.0330	97.625	0.01024	0.31024	3.223
14	39.374	0.0254	127.913	0.00782	0.30782	3.249
15	51.186	0.0195	167.286	0.00598	0.30598	3.268
16	66.542	0.0150	218.472	0.00458	0.30458	3.283
17	86.504	0.0116	285.014	0.00351	0.30351	3.295
18	112.455	0.0089	371.518	0.00269	0.30269	3.304
19	146.192	0.0068	483.973	0.00207	0.30207	3.311
20	190.050	0.0053	630.165	0.00159	0.30159	3.316
21	247.065	0.0040	820.215	0.00122	0.30122	3.320
22	321.184	0.0031	1067.280	0.00094	0.30094	3.323
23	417.539	0.0024	1388.464	0.00072	0.30072	3.325
24	542.801	0.0018	1806.003	0.00055	0.30055	3.327
25	705.641	0.0014	2348.803	0.00043	0.30043	3.329
26	917.333	0.0011	3054.444	0.00033	0.30033	3.330
27	1192.533	0.0008	3971.778	0.00025	0.30025	3.331
28	1550.293	0.0006	5164.311	0.00019	0.30019	3.331
29	2015.381	0.0005	6714.604	0.00015	0.30015	3.332
30	2619.996	0.0004	8729.985	0.00011	0.30011	3.332
35	9727.860	0.0001	32422.868	0.00003	0.30003	3.333

Table 19-8. 40% Interest Factors

Period n	Single-payment compound-amount (SPCA) Future value of $1 $(1 + i)^n$	Single-payment present-worth (SPPW) Present value of $1 $\dfrac{1}{(1 + i)^n}$	Uniform-series compound-amount (USCA) Future value of uniform series of $1 $\dfrac{(1 + i)^n - 1}{i}$	Sinking-fund payment (SFP) Uniform series whose future value is $1 $\dfrac{i}{(1 + i)^n - 1}$	Capital recovery (CR) Uniform series with present value of $1 $\dfrac{i(1 + i)^n}{(1 + i)^n - 1}$	Uniform-series present-worth (USPW) Present value of uniform series of $1 $\dfrac{(1 + i)^n - 1}{i(1 + i)^n}$
1	1.400	0.7143	1.000	1.00000	1.40000	0.714
2	1.960	0.5102	2.400	0.41667	0.81667	1.224
3	2.744	0.3644	4.360	0.22936	0.62936	1.589
4	3.842	0.2603	7.104	0.14077	0.54077	1.849
5	5.378	0.1859	10.946	0.09136	0.49136	2.035
6	7.530	0.1328	16.324	0.06126	0.46126	2.168
7	10.541	0.0949	23.853	0.04192	0.44192	2.263
8	14.758	0.0678	34.395	0.02907	0.42907	2.331
9	20.661	0.0484	49.153	0.02034	0.42034	2.379
10	28.925	0.0346	69.814	0.01432	0.41432	2.414
11	40.496	0.0247	98.739	0.01013	0.41013	2.438
12	56.694	0.0176	139.235	0.00718	0.40718	2.456
13	79.371	0.0126	195.929	0.00510	0.40510	2.469
14	111.120	0.0090	275.300	0.00363	0.40363	2.478
15	155.568	0.0064	386.420	0.00259	0.40259	2.484
16	217.795	0.0046	541.983	0.00185	0.40185	2.489
17	304.913	0.0033	759.784	0.00132	0.40132	2.492
18	426.879	0.0023	1064.697	0.00094	0.40094	2.494
19	597.630	0.0017	1491.576	0.00067	0.40067	2.496
20	836.683	0.0012	2089.206	0.00048	0.40048	2.497
21	1171.356	0.0009	2925.889	0.00034	0.40034	2.498
22	1639.898	0.0006	4097.245	0.00024	0.40024	2.498
23	2295.857	0.0004	5737.142	0.00017	0.40017	2.499
24	3214.200	0.0003	8032.999	0.00012	0.40012	2.499
25	4499.880	0.0002	11247.199	0.00009	0.40009	2.499
26	6299.831	0.0002	15747.079	0.00006	0.40006	2.500
27	8819.764	0.0001	22046.910	0.00005	0.40005	2.500

Table 19-9. 50% Interest Factors

Period n	Single-payment compound-amount (SPCA) Future value of $1 $(1 + i)^n$	Single-payment present-worth (SPPW) Present value of $1 $\dfrac{1}{(1 + i)^n}$	Uniform-series compound-amount (USCA) Future value of uniform series of $1 $\dfrac{(1 + i)^n - 1}{i}$	Sinking-fund payment (SFP) Uniform series whose future value is $1 $\dfrac{i}{(1 + i)^n - 1}$	Capital recovery (CR) Uniform series with present value of $1 $\dfrac{i(1 + i)^n}{(1 + i)^n - 1}$	Uniform-series present-worth (USPW) Present value of uniform series of $1 $\dfrac{(1 + i)^n - 1}{i(1 + i)^n}$
1	1.500	0.6667	1.000	1.00000	1.50000	0.667
2	2.250	0.4444	2.500	0.40000	0.90000	1.111
3	3.375	0.2963	4.750	0.21053	0.71053	1.407
4	5.062	0.1975	8.125	0.12308	0.62308	1.605
5	7.594	0.1317	13.188	0.07583	0.57583	1.737
6	11.391	0.0878	20.781	0.04812	0.54812	1.824
7	17.086	0.0585	32.172	0.03108	0.53108	1.883
8	25.629	0.0390	49.258	0.02030	0.52030	1.922
9	38.443	0.0260	74.887	0.01335	0.51335	1.948
10	57.665	0.0173	113.330	0.00882	0.50882	1.965
11	86.498	0.0116	170.995	0.00585	0.50585	1.977
12	129.746	0.0077	257.493	0.00388	0.50388	1.985
13	194.620	0.0051	387.239	0.00258	0.50258	1.990
14	291.929	0.0034	581.859	0.00172	0.50172	1.993
15	437.894	0.0023	873.788	0.00114	0.50114	1.995
16	656.841	0.0015	1311.682	0.00076	0.50076	1.997
17	985.261	0.0010	1968.523	0.00051	0.50051	1.998
18	1477.892	0.0007	2953.784	0.00034	0.50034	1.999
19	2216.838	0.0005	4431.676	0.00023	0.50023	1.999
20	3325.257	0.0003	6648.513	0.00015	0.50015	1.999
21	4987.885	0.0002	9973.770	0.00010	0.50010	2.000
22	7481.828	0.0001	14961.655	0.00007	0.50007	2.000

Table 19-10. 5 Year Escalation Table

Present Worth of a Series of Escalating Payments Compounded Annually
Discount-Escalation Factors for $n = 5$ Years

Discount Rate	Annual Escalation Rate					
	0.10	0.12	0.14	0.16	0.18	0.20
0.10	5.000000	5.279234	5.572605	5.880105	6.202627	6.540569
0.11	4.866862	5.136200	5.420152	5.717603	6.029313	6.355882
0.12	4.738562	5.000000	5.274242	5.561868	5.863289	6.179066
0.13	4.615647	4.869164	5.133876	5.412404	5.704137	6.009541
0.14	4.497670	4.742953	5.000000	5.269208	5.551563	5.847029
0.15	4.384494	4.622149	4.871228	5.131703	5.404955	5.691165
0.16	4.275647	4.505953	4.747390	5.000000	5.264441	5.541511
0.17	4.171042	4.394428	4.628438	4.873599	5.129353	5.397964
0.18	4.070432	4.287089	4.513947	4.751566	5.000000	5.259749
0.19	3.973684	4.183921	4.403996	4.634350	4.875619	5.126925
0.20	3.880510	4.084577	4.298207	4.521178	4.755725	5.000000
0.21	3.790801	3.989001	4.196400	4.413341	4.640260	4.877689
0.22	4.704368	3.896891	4.098287	4.308947	4.529298	4.759649
0.23	3.621094	3.808179	4.003835	4.208479	4.422339	4.645864
0.24	3.540773	3.722628	3.912807	4.111612	4.319417	4.536517
0.25	3.463301	3.640161	3.825008	4.018249	4.220158	4.431144
0.26	3.388553	3.560586	3.740376	3.928286	4.124553	4.329514
0.27	3.316408	3.483803	3.658706	3.841442	4.032275	4.231583
0.28	3.246718	3.409649	3.579870	3.757639	3.943295	4.137057
0.29	3.179393	3.338051	3.503722	3.676771	3.857370	4.045902
0.30	3.114338	3.268861	3.430201	3.598653	3.774459	3.957921
0.31	3.051452	3.201978	3.359143	3.523171	3.694328	3.872901
0.32	2.990618	3.137327	3.290436	3.450224	3.616936	3.790808
0.33	2.931764	3.074780	3.224015	3.379722	3.542100	3.711472
0.34	2.874812	3.014281	3.159770	3.311524	3.469775	3.634758

Table 19-11. 10 Year Escalation Table

Present Worth of a Series of Escalating Payments Compounded Annually
Discount-Escalation Factors for *n* = 10 Years

Discount Rate	Annual Escalation Rate					
	0.10	0.12	0.14	0.16	0.18	0.20
0.10	10.000000	11.056250	12.234870	13.548650	15.013550	16.646080
0.11	9.518405	10.508020	11.613440	12.844310	14.215140	15.741560
0.12	9.068870	10.000000	11.036530	12.190470	13.474590	14.903510
0.13	8.650280	9.526666	10.498990	11.582430	12.786980	14.125780
0.14	8.259741	9.084209	10.000000	11.017130	12.147890	13.403480
0.15	7.895187	8.672058	9.534301	10.490510	11.552670	12.731900
0.16	7.554141	8.286779	9.099380	10.000000	10.998720	12.106600
0.17	7.234974	7.926784	8.693151	9.542653	10.481740	11.524400
0.18	6.935890	7.589595	8.312960	9.113885	10.000000	10.980620
0.19	6.655455	7.273785	7.957330	8.713262	9.549790	10.472990
0.20	6.392080	6.977461	7.624072	8.338518	9.128122	10.000000
0.21	6.144593	6.699373	7.311519	7.987156	8.733109	9.557141
0.22	5.911755	6.437922	7.017915	7.657542	8.363208	9.141752
0.23	5.692557	6.192047	6.742093	7.348193	8.015993	8.752133
0.24	5.485921	5.960481	6.482632	7.057347	7.690163	8.387045
0.25	5.290990	5.742294	6.238276	6.783767	7.383800	8.044173
0.26	5.106956	5.536463	6.008083	6.526298	7.095769	7.721807
0.27	4.933045	5.342146	5.790929	6.283557	6.824442	7.418647
0.28	4.768518	5.158489	5.585917	6.054608	6.568835	7.133100
0.29	4.612762	4.984826	5.392166	5.838531	6.327682	6.864109
0.30	4.465205	4.820429	5.209000	5.634354	6.100129	6.610435
0.31	4.325286	4.664669	5.035615	5.441257	5.885058	6.370867
0.32	4.192478	4.517015	4.871346	5.258512	5.681746	6.144601
0.33	4.066339	4.376884	4.715648	5.085461	5.489304	5.930659
0.34	3.946452	4.243845	4.567942	4.921409	5.307107	5.728189

Table 19-12. 15 Year Escalation Table

	Present Worth of a Series of Escalating Payments Compounded Annually Discount-Escalation Factors for n = 15 years					
Discount Rate	Annual Escalation Rate					
	0.10	0.12	0.14	0.16	0.18	0.20
0.10	15.000000	17.377880	20.199780	23.549540	27.529640	32.259620
0.11	13.964150	16.126230	18.690120	21.727370	25.328490	29.601330
0.12	13.026090	15.000000	17.332040	20.090360	23.355070	27.221890
0.13	12.177030	13.981710	16.105770	18.616160	21.581750	25.087260
0.14	11.406510	13.057790	15.000000	17.287320	19.985530	23.169060
0.15	10.706220	12.220570	13.998120	16.086500	18.545150	21.442230
0.16	10.068030	11.459170	13.388900	15.000000	17.244580	19.884420
0.17	9.485654	10.766180	12.262790	14.015480	16.066830	18.477610
0.18	8.953083	10.133630	11.510270	13.118840	15.000000	17.203010
0.19	8.465335	9.555676	10.824310	12.303300	14.030830	16.047480
0.20	8.017635	9.026333	10.197550	11.560150	13.148090	15.000000
0.21	7.606115	8.540965	9.623969	10.881130	12.343120	14.046400
0.22	7.227109	8.094845	9.097863	10.259820	11.608480	13.176250
0.23	6.877548	7.684317	8.614813	9.690559	10.936240	12.381480
0.24	6.554501	7.305762	8.170423	9.167798	10.320590	11.655310
0.25	6.255518	6.956243	7.760848	8.687104	9.755424	10.990130
0.26	5.978393	6.632936	7.382943	8.244519	9.236152	10.379760
0.27	5.721101	6.333429	7.033547	7.836080	8.757889	9.819020
0.28	5.481814	6.055485	6.710042	7.458700	8.316982	9.302823
0.29	5.258970	5.797236	6.410005	7.109541	7.909701	8.827153
0.30	5.051153	5.556882	6.131433	6.785917	7.533113	8.388091
0.31	4.857052	5.332839	5.872303	6.485500	7.184156	7.982019
0.32	4.675478	5.123753	5.630905	6.206250	6.860492	7.606122
0.33	4.505413	4.928297	5.405771	5.946343	6.559743	7.257569
0.34	4.345926	4.745399	5.195502	5.704048	6.280019	6.933897

Table 19-13. 20 Year Escalation Table

Present Worth of a Series of Escalating Payments Compounded Annually
Discount-Escalation Factors for n = 20 Years

Discount Rate	Annual Escalation Rate					
	0.10	0.12	0.14	0.16	0.18	0.20
0.10	20.000000	24.295450	29.722090	36.592170	45.308970	56.383330
0.11	18.213210	22.002090	26.776150	32.799710	40.417480	50.067940
0.12	16.642370	20.000000	24.210030	29.505400	36.181240	44.614710
0.13	15.259850	18.243100	21.964990	26.634490	32.502270	39.891400
0.14	14.038630	16.694830	20.000000	24.127100	29.298170	35.789680
0.15	12.957040	15.329770	18.271200	21.929940	26.498510	32.218060
0.16	11.995640	14.121040	16.746150	20.000000	24.047720	29.098950
0.17	11.138940	13.048560	15.397670	18.300390	21.894660	26.369210
0.18	10.373120	12.093400	14.201180	16.795710	20.000000	23.970940
0.19	9.686791	11.240870	13.137510	15.463070	18.326720	21.860120
0.20	9.069737	10.477430	12.188860	14.279470	16.844020	20.000000
0.21	8.513605	9.792256	11.340570	13.224610	15.527270	18.353210
0.22	8.010912	9.175267	10.579620	12.282120	14.355520	16.890730
0.23	7.555427	8.618459	9.895583	11.438060	13.309280	15.589300
0.24	7.141531	8.114476	9.278916	10.679810	12.373300	14.429370
0.25	6.764528	7.657278	8.721467	9.997057	11.533310	13.392180
0.26	6.420316	7.241402	8.216490	9.380883	10.778020	12.462340
0.27	6.105252	6.862203	7.757722	8.823063	10.096710	11.626890
0.28	5.816151	6.515563	7.339966	8.316995	9.480940	10.874120
0.29	5.550301	6.198027	6.958601	7.856833	8.922847	10.194520
0.30	5.305312	5.906440	6.609778	7.437339	8.416060	9.579437
0.31	5.079039	5.638064	6.289875	7.054007	7.954518	9.021190
0.32	4.869585	5.390575	5.995840	6.702967	7.533406	8.513612
0.33	4.675331	5.161809	5.725066	6.380829	7.148198	8.050965
0.34	4.494838	4.949990	5.475180	6.084525	6.795200	7.628322

20

Case Study 1: Computing EMC System Energy Savings for a New Commercial Building

Delbert M. Fowler, *P.E.*
PS Consultants, Inc.

The energy savings computations discussed in this chapter* were done for a 49 floor all-electric high rise office building of 1.4 million gross square feet in downtown Dallas. The building exterior consists of double pane reflective glass, 37% of which is vision glass. The cooling system includes two 1520 ton chillers and one 760 ton chiller, serving variable air volume air handlers every other floor. Heating is by means of resistance duct heaters and fan-powered mixing boxes with electric heating coils downstream.

The owner, a major developer, needed assistance in deciding whether or not to install an energy management and control system. They also wanted assistance in deciding which loads were economical to control if they decided to install an EMCS.

As is normally the case for a new building, the basic building design was essentially complete when these studies were started. For each separate computation below, the local controls as provided in the initial design documents and specifications are outlined, followed

---

*This case study was first presented as a report to the Fifth World Energy Engineering Congress, when Mr. Fowler was a principal with Fowler/Blum Energy Consultants, Inc.

by a brief discussion and the EMCS energy savings calculations. Before this energy savings study was undertaken, a computer building simulation study had been completed for the purpose of establishing the design energy budget or goal. This earlier study will be referred to from time to time.

POSSIBLE SAVINGS

Energy Use

A similar existing all-electric high rise office building in Dallas with double-pane reflective glass that is operated on a rigid schedule of 12 hours per day and 6 hours for Saturday of HVAC uses energy at the rate of 87,000 Btu/sq ft/yr, or 25.49 KWH/sq ft/yr. It does not have a microprocessor or computer-based energy management control system installed, but the building operator does manually adjust the chilled water temperature with the outdoor weather conditions. Lighting is 4 watts/sq ft, whereas previous energy studies of the new building used 2 watts/sq ft as the lighting level.

If the 4 watts/sq ft of lighting is reduced to 2 watts/sq foot, the existing building energy would be reduced to 17 KWH/sq ft/yr, or 58,000 Btu/sq ft/yr. Such an actual energy operating budget is appropriate for consideration for the new building as a conservative estimate on which to calculate savings. Annual energy costs (KWH and fuel costs) at current rates (DP&L Rate G) amount to $665,700.

$$17 \text{ KWH} \times 1,412,657 \text{ GSF} \times \$.02772 = \$665,700/\text{Year}$$

For demand use, the monthly peak demand calculations from the first energy study of the new building are applied. These demand use figures are for electric cooling with centrifugal machines and for electric resistance heat.

Cooling months of Jun, Jul, Aug, Sep, Oct (DP&L Rate G)

Month	Demand	Cost (@ $7.56)
Jun	7462	$ 56,412.72
Jul	7461	56,405.16
Aug	7464	56,427.84
Sep	7001	52,927.56
Oct	7083	53,547.48
TOTAL		$275,720.76

Heating months of Nov, Dec, Jan, Feb, Mar, Apr (DP&L Rate GH electric heating rider)

Billed demand (winter months) = (.5625 × *A*) + (.25 × *B*)

where *A* = Summer peak demand (Aug = 7465)

   B = Actual or recorded demand

Month	Actual Demand	Billed Demand	Cost (@ $7.56)
Nov	6907	5925	$ 44,793.00
Dec	6593	5847	44,203.32
Jan	6591	5946	44,951.76
Feb	6229	5756	43,515.36
Mar	5985	5695	43,054.20
Apr	5745	5635	42,600.60
		TOTAL	$ 263,118.24
The month of May had a demand of 6671			50,432.76
Yearly Demand Total			589,271.76
Yearly Energy & Fuel Total			665,700.00
		TOTAL	$1,254,972.00

This total amounts to an annual cost of 89¢/sq ft for the 1,412,657 gross square feet.

Energy Savings

Energy Management and Control Systems in general will save 10–15% of annual electricity costs when only start/stop scheduling, optimized start/stop, duty cycling, demand monitoring and limiting, and night set-back are used. In other words, these savings result without either lighting control, chilled water reset, condenser water reset, or chiller optimization. But these are savings which normally result in existing buildings even where the building operators have been very energy conscious. These systems can improve upon normal manual operation and can save on demand charges where manual or time-clock operation can not.

	Annual	*10% Saving*	*15% Saving*
Energy and Fuel Charges	$ 665,700	$ 66,570	$ 99,855
Demand Charges	$ 589,272	$ 58,927	$ 87,941
Total	$1,254,972	$125,497	$187,796

Pay-Back

A pay-back criteria of three years would permit an economic expenditure of three times the above savings.

$$3 \times \$125,497 = \$376,491$$
$$3 \times \$187,796 = \$563,388$$

DETAILED SAVINGS CALCULATIONS

Since there are approximately 50 air handlers in the building, and since it is necessary to control each one of these for after-hours usage at nights and on week-ends, a microprocessor which can handle a maximum number of 40 loads is adequate. Savings estimates which follow, will be made only for microcomputer or computer-based systems.

In each of the savings estimates which follows, the local controls as currently specified are described, followed by a discussion of these local controls and how an EMCS would function to save energy. The discussion is followed by the energy savings estimated with installation of a computer-based EMCS.

HEATING, VENTILATING AND AIR CONDITIONING (HVAC) AIR SIDE SYSTEMS

Start/Stop of Air Handling Units

Local Controls: A single 10-step cam action 7-day time clock on the Engineer's Central Control Panel controls office space air hanling units. Ventilating units, exhaust fans, and the like are controlled manually. Four outside air supply units are interlocked with the air handling units which they serve. Overtime use is enabled by a courtesy panel for floors one through four and by manual override above the fourth floor. A second time clock is provided for chiller plant equipment.

Discussion: Time clock control is adequate for a single tenant building of smaller size with reasonably fixed operating hours. With time clocks, air handlers can be started at the appropriate time in the morning, but must run until the latest tenant leaves in the evening or must be enabled by a manual override if all are turned off at the end of the day. The manual override system does not permit the tenant to be easily billed for after-hours usage, unless someone from the operating staff performs the override function. For 50 air handlers this does not seem practical from a personnel and economic viewpoint.

A microprocessor is not suited for this building because it is limited to 40 loads, whereas this building requires a capability for 50 just for the air handlers.

Energy Savings: If all fans were permitted to run until the last tenant and the janitors were completed with their work, say at 10 PM, an additional four hours of run time would be needed over what would be required if the fans could be controlled individually by an EMCS. If the fans are all turned off at 6 PM and tenants and janitorial crew have access to a manual override, at least two hours of additional run time would be involved.

A saving then of 2–4 hours should be possible for the 1000 KW of fans, with electricity cost at $0.02772 and no demand saving. Since the variable air volume fans will be operating under part load conditions during these periods, a load factor of .6 will be used for them.

(2–4 hours) (5 days/week) (52 weeks/year) (1000 KW) (.60)
($.02772/KWH) = $8,649 − $17,298

"Optimum" Start/Stop and "Set Up" Temperature

Local Controls: The single time clock on the Engineer's Central Control Panel can be adjusted to start units at a fixed time before occupancy to permit pre-cooling of the building. During the heating season, the fan-powered mixing boxes remain on around the clock to maintain minimum occupancy temperatures.

Discussion: This program saves energy by starting the heating or cooling system only as early as is necessary to achieve desired indoor comfort conditions, with the start time based on either outside or inside temperatures.

Energy Savings: For the 1000 KW of fans, the optimum start/stop program will save an estimated one to two hours for each operating day of the year, for a savings of $4,324–$8,648.

$$(1 \text{ hour}) (5) (52) (1000) (.6) (\$.02772) = \$4324$$

**Control of Heating—
Night Set-Back and
Summer Lock-Out
of the Heating System**

Local Controls: The building is heated by electric resistance coils in the fan-powered mixing boxes (FPMB). Heating is controlled in sequence by the same pneumatic thermostat system which controls the cooling. The FPMBs will cycle on at night during the heating season to maintain the temperature set point of the thermostats; and unless disabled, they will also cycle on in similar fashion during the cooling season.

Discussion: There are about 500 heating units in the building, so control of them by an EMCS is costly in terms of the relays and number of points involved. But the savings are correspondingly high. The fan-powered mixing boxes are circuited electrically separately from the electric heating coils. This allows both to be controlled separately.

Energy Savings: Energy can be saved through a form of night set-back and through locking out the FPMBs during the cooling season.

Night Set-Back and Summer Lock-Out Heating Savings: Savings here can be achieved by locking out heating above a fixed outside temperature, or inside temperatures if an adequate number of temperature sensors are provided in the EMCS. An estimated 1/6 to 1/3 of required building heating can be saved in this manner. From our earlier computer building simulation, 4,439,622 KWH of heat were required for the year.

$$(1/6-1/3) (4,439,622) (\$.02772) = \$20,511-\$41,022$$

Summer Lock-Out of the Heating System Fan Savings: The fans in the 500 heating units (295 KW) could be locked out for an estimated 200 hours during the cooling season for an annual savings of $16,355.

$$295 \times 2000 \times \$.02772 = \$16,355$$

Outside Air Control—
Supply and Exhaust

Local Controls: Four outside air supply units are interlocked at the Engineer's Central Control Panel with the air handling units which they serve. Inlet vanes on each unit are controlled by static pressure. Exhaust fans are not tied in to the Engineer's Central Control Panel.

Discussion—Supply Air Fans: The supply air fans are not required except during occupied hours. The extra hours of operation during morning cool down of the building are useful only if the outside air is assisting in the cooling of the building. At least four hours per day of operating time for the supply fans can be saved.

Cooling Requirements

Wet Bulb °F*	Enthalpy Btu/#	Enthalpy to 63° WB (28.5 Btu/#)	#Hours per Year*	Btu/# Dry Air Hours per Year
75	38.6	10.1	54	545
74	37.7	9.2	229	2,107
73	36.8	8.3	404	3,353
72	35.9	7.4	566	4,188
70	34.1	5.6	808	4,525
68	32.4	3.9	995	3,880
65	30.0	1.5	966	1,449
				19,947

*Engineering Weather Data NAVFAC P-89 1 July 1978

Energy Savings:

Cooling

19,947 Btu/#Hours × 103,600 CFM ×

$$\frac{60 \text{ Min./Hr.}}{13.5 \text{ CF/\#}} \times \frac{4 \text{ Hrs}}{24 \text{ Hrs}} \times \frac{5 \times 52}{365 \text{ Days}}$$
= 1,090 Million Btu/Year

$$\frac{1,090 \text{ Mil. Btu/Yr.}}{12,000 \text{ Btu/Ton Hr.}} \times 1 \text{ KWH/Ton Hour} \times \$0.02772/\text{KWH}$$

= $2,519/ Year

Fans

4 Hours/Day × 260 Days × 103,600 CFM ×

$$\frac{4.5" \text{ S.P.}}{5745} \times \$.02772 = \$2,399$$

Total Annual Savings = $4,858

Discussion—Exhaust Air Fans: Exhaust air fans normally operate 24 hours per day and usually are not controlled in any way. If controlled by the EMCS they can be turned off for at least 10 hours of each operating day and 24 hours for each nonoperating day.

$$10 \times 5 \times 52 = 2600 \text{ Hours}$$
$$24 \times 2 \times 52 = \underline{2496 \text{ Hours}}$$
$$5096 \text{ Hours}$$

Energy Savings:

Cooling

$$\frac{5096 \text{ Hours}}{8760 \text{ Hours}} \times 23,500 \text{ CFM of controllable exhaust}$$

× 19,947 Btu/# Dry Air ×

$$\frac{60 \text{ Min./Hr}}{13.5 \text{ CFM}} \quad \frac{1 \text{ KWH/Ton Hr}}{12,000 \text{ Btu/Ton Hr}}$$
= 100,997 KWH/Year

Heating

23,500 CFM X 2,400 Heating Degree Days X 24 Hrs/Day X
1.08 Btu/CFM °F Hour X

$$\frac{5096 \text{ Hours}}{8760 \text{ Hours } 3413 \text{ Btuh/KWH}} = 249,174 \text{ KWH/Year}$$

Fans

$$\frac{23,500 \text{ CFM X } 1"}{5745} \text{ X } 5096 \text{ Hours} = 20,845 \text{ KWH/Year}$$

Total Annual Energy Saving 371,016 KWH/Year
    371,016 X $.02772 = $10,285/Year

CHILLER CONTROL

Local Controls: Lead or first chiller with chilled water pump, condenser water pump, and cooling tower is started with a time clock.
Additional chillers are started or stopped manually by the building
operators. Building operators normally leave the chilled water temperature at the design setting. Condenser water temperature is controlled to the lowest permitted by the chiller manufacturer.

Discussion: The building operator will have to start and stop
chillers for both day and evening use, based on the judgment of the
operator. Specific starting and stopping sequences in an EMCS program can save a large number of hours of chiller operation, compared
to manual procedures. Many building operators begin the day by automatically starting 2 or 3 chillers at the same time on the hottest
summer days and let them run all day. So the start/stop/selection
procedure is important for savings.

Second, building operators normally do not change the leaving
chilled water temperature of each chiller, even though that temperature is only required for the few days of the year that meet or exceed design conditions. An EMCS can continually adjust the leaving
water temperature, for a saving of over 1% for each degree the temperature is raised.

Third, condenser water temperature can be continually reset by an
EMCS in the same manner as the chilled water temperature, with a

similar saving. An EMCS can also reset the condenser water temperature by the ambient or outside wet bulb temperature and cycle off tower fans as the condenser water temperature approaches the outside wet bulb temperature, saving additional fan horsepower.

Fourth, an EMCS can be designed to provide demand control for chillers. In addition to the savings possible through proper chiller selection to meet the instantaneous load, as already discussed, the demand limiter on each chiller can be used to set up digital outputs for stepped load shedding of each chiller.

Automatic chiller start/stop controls are available from Carrier, Chillitrol, CESCO, Johnson, Barber–Colman, and others, in addition to being available as a part of EMC systems. Some controls are limited to handling one chiller and most can handle only two. The logic options in handling more than two obviously get more detailed.

Energy Savings:

Chiller Start/Stop: Experience indicates that a good building operator will start and leave on the 2nd and 3rd chillers 25% more time than is necessary with automatic controls. And from our earlier computer building simulation, we estimate that chiller #2 (1520 tons) will need to run 1500 hours and chiller #3 (760 tons) for 500 hours. The EMCS would be able to save the added 25%, which we estimate would be at a minimum energy input of 20% of peak chiller input.

Chiller #2

(1520 tons) (.8 KW/ton) (1500 Hours) (.25) (.2) =   91,200 KWH
CHW and CW Pumps
   74 KW × 1500 × .25   =   27,750 KWH
                                                              118,950 KWH

Chiller #3

(760 tons) (.8 KW/ton) (500 Hours) (.25) (.2)   =   15,200 KWH
CHW and CW Pumps
   37 KW × 500 × .25   =   4,625 KWH
                                                             29,825 KWH

TOTAL = 138,775 KWH

Savings = (138,775) ($.02772/KWH)   = $3,847

Chilled Water Reset: From our earlier computer building simulation we find that chillers for the new building will use 2,965,715 KWH per year. For this savings calculation, we assume from experience a weighted average chilled water increase for the entire year of 6° for a 6% saving.

$$2,965,715 \times .06 \times \$.02772 = \$4,933$$

Cooling Tower: Based on experience, we assume that cooling tower fans can be cycled off when the condenser water temperature is within 3° of the outside wet bulb temperature. We estimate that this can be done 50% of the time, when chillers are less than 40% loaded.

$$(2000 \text{ Hours}) (.5) (180 \text{ KW}) \times \$.02772 = \$4,990$$

Total energy savings then for chiller control are, very conservatively:

Automatic Start/Stop/Selection	$ 3,847
Chilled Water Reset	4,933
Cooling Tower	4,990
	$13,770

DEMAND MONITORING AND CONTROL

Local Controls: There is no way to provide demand monitoring and control or limiting through local controls. This function can only be provided through a microprocessor or computer-based EMCS.

Discussion: Peak demands may be classified as being of three general types: *Morning start-up* (Monday morning is usually the most pronounced, if the building cooling system has been shut down over the weekend); *Daytime peaks* and *Random peaks.* The manner in which these may appear on a graphical demand meter are shown in the example on the next page.

Morning Start-Up: The only effective way of controlling this type of peak is through use of some type of chiller control system which provides for a *soft start.* Among these types, of course, is that provided by a computer-based EMCS.

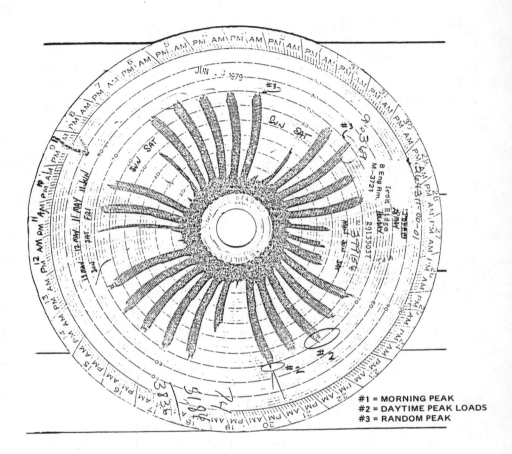

#1 = MORNING PEAK
#2 = DAYTIME PEAK LOADS
#3 = RANDOM PEAK

Daytime Peaks: These peaks are caused by the natural simultaneous solar, transmission, outside air and internal load peaks and may last for several hours. In large office buildings, these peaks can be controlled only by limiting chiller loading and allowing some drift upward in space temperature.

It should be recognized that when a cooling system is designed and sized based on *1% design weather conditions,* this means that, on the average, there will be only 30 hours during the year that are above these conditions. For the Dallas area the 1% summer design conditions include a 102°F dry bulb temperature with a mean coincident wet bulb temperature of 75°F.

Random Peaks: These peaks occur in most buildings and are caused by simultaneous operation of miscellaneous equipment: hot water heaters, fans, elevators, and the like. These peaks can be controlled by an EMCS with demand monitoring and load shedding. Often a load will have to be shed for only one 15 minute demand interval. If the load to be shed is one step on a chiller, there will not be any discernible change in comfort conditions for such a short period of time.

All the earlier energy savings calculations were based on 1981 rates for energy (KWH) and fuel (per KWH), plus 5% sales tax. The figure of $.02772/KWH includes no demand increment.

Summer Demand Savings

The loads that would be shed for summer demand limiting are the small chiller, one step of 20% on one of the two large chillers, and the electric domestic hot water heaters.

small chiller 608 KW + 37 KW of Pumps	=	645 KW
large chiller (20%) 1216 KW × .2	=	243 KW
hot water heaters 96 KW × .5 (diversity of .5)	=	48 KW
		936 KW

With proper setting of demand targets for each month, it should be possible to save 936 KW for each of the summer months (5) and May.

$$936 \times 6 \times \$7.56 = \$42,457$$

And since the mathematics of the rate structure provide a minimum bill or ratchet, there is also a saving for the six winter months.

Winter billing demand = .5625 summer peak + .25 actual winter

Thus, .5625 × 926 or 526.5 KW of demand costs are avoided for the winter months.

$$526.5 \text{ KW} \times 6 \times \$7.56 = \$23,882$$

A reduction of 936 KW for each of the five summer cooling months plus May, results in an annual saving of $66,339.

Winter Demand Savings

The heating rider GH is in effect for this electrically heated building. The saving of demand costs is accordingly less than in the summer months, since the actual demand for the heating months is reduced. From the formula above, we see that winter billing demand is reduced only by an amount equal to 1/4 or .25 of each KW of actual demand for the billing period.

The loads which would be shed for the winter months include 40% of the small chiller (or the equivalent if a large one is operating), the electric hot water heaters, and 10 floors of electric heat for 12 minutes each hour.

$$
\begin{array}{llr}
\text{small chiller (40\%)} & 608 \times .4 = & 243 \text{ KW} \\
\text{hot water heaters} & 96 \times .5 = & 48 \text{ KW} \\
\text{10 floors @ 55 KW each} & = & \underline{550 \text{ KW}} \\
\text{TOTAL} & = & 841 \text{ KW}
\end{array}
$$

$$841 \text{ KW} \times .25 \times 6 \times \$7.56 = \$9,537$$

Total annual demand cost savings then amount to $75,876.

ELECTRIC DOMESTIC HOT WATER HEATERS

Local Controls: The 16 electric hot water heaters (52 gallons and 6 KW each) and 16 1/12-HP circulating pumps make up the hot water system. Each normally cycles on as required to maintain a set leaving water temperature, and the circulating pump runs continuously unless turned off.

Discussion: Controlling these water heaters for energy savings alone is not economical. But when demand savings, discussed earlier, are included, such control does become feasible.

LIGHTING CONTROL

Local Controls: The basic design involves a modular wiring system with local switching in individual offices and floor control, if desired, using the regular circuit breakers.

A bid alternate provided low voltage relays for each pair of lighting circuits (one per 1000 square feet of floor space, approximately) in the event connection with an energy management system would be required. The contractor priced this alternate at $166,170.

Discussion: Ideally, lighting circuits should be:

- "enabled" just prior to the first occupant arriving.

- switched on in each individual office or space by occupant of that office when and if they arrive for work.

- switched off in each individual office or space by the occupant at any time during the day that the occupant(s) leave.

- "disabled" as soon as all occupants have departed.

- "enabled" a floor at a time by the night-time cleaning crew when they begin cleaning that floor.

- "disabled" by the night-time cleaning crew when they complete their cleaning on a floor.

With currently provided local controls, those actions entitled "enable" and "disable" involve the switching of floor circuit breakers in the electrical closet on each floor. Switching of these circuit breakers on 50 floors is obviously a time consuming task, even if only two trips a day by building operators or security personnel are necessary.

The provision of local switching provides added energy savings, particularly in perimeter offices and unoccupied offices. A recent article in the Illuminating Engineering Society (IES) magazine stated that energy savings in lighting of as much as 40% and demand savings as much as 15% resulted when local switching was added where only floor circuit breakers had existed. Thus, any good energy saving lighting system should provide local switching to enable the occupants to effect these savings.

With the bid alternate low voltage relays, a computer-based EMCS can perform those functions above defined as "enable" and "disable."

Energy Savings: With the ability to "enable" a floor remotely through the EMCS instead of having a building security employee do so at the floor circuit breaker, a major saving is possible in hours or use. Whereas the hours of use with circuit breaker control may be

7 AM–9 PM, with EMCS remote control, it should be possible to reduce the hours of use of "on time" to 7 AM–6 PM or 7AM–7 PM. Anyone needing light after 6 PM can call the central after-hours security office on the first floor and have the lights for that floor "enabled" through the EMCS for a specified period.

Such a 3 hour saving for the 2560 KW of lighting in the building would save

$$3 \text{ hours} \times 5 \text{ days/week} \times 52 \text{ weeks/year} \times 2560 \text{ KW}$$
$$= 1,996,800 \text{ KWH}$$

When lighting loads are reduced, cooling loads are also reduced, and heating requirements are increased since the heat of light helps to heat the building.

The added heating requirement will be that heat of light which must be replaced by resistance heat for the heating season of 15 weeks.

$$3 \times 5 \times 15 \times 2560 = 576,000 \text{ KWH}$$

The added cooling saving will result from the reduced heat of light which does not have to be removed for the cooling season of 35 weeks.

$$\frac{3 \times 5 \times 35 \times 2560}{12,000 \text{ Btu/Ton-Hour}} \times 3413 \text{ Btu/KWH}$$
$$\times 1.0 \text{ KWH/Ton-Hour} = 382,256 \text{ KWH}$$

The net energy saving is 1,803,056 KWH, which at a cost of $.02772 is equivalent to $49,981 per year. If the daily saving amounted to only 2 hours, the annual saving would be $33,320.

Payback: If the 50 points for lighting control cost $500 per point, the total cost of lighting control will be $191,170.

$$
\begin{array}{lll}
\$500 \times 50 & = \$ \ 25,000 \\
\text{Bid Alternate} & = \underline{\ \ 166,170} \\
& \ \ \ \$191,170
\end{array}
$$

The payback is less than four years if 3 hours of lighting use are saved each day, and over 5 years if only 2 hours are saved.

Future Savings: If the bid alternate relays are purchased now with single point control (one digital output per floor), there are future potential savings to be had.

Each of the 50 floors is now designed for about 6 circuits for the common core areas and 36 circuits for the tenant areas. Let us assume that each floor of the building will average 3 hours of after-hours light use per day, or 18 hours per week, and that only two of the 36 tenant circuits are required for that use. Each circuit averages about 1.5 KW.

If the EMCS is converted at a later date by installing a multiplexer on each floor so that each relay (2 circuits) may be controlled individually, only 2 circuits would need to be "enabled" after hours, instead of 36. The energy for 34 circuits would be saved, for an annual saving of $66,162.

$$34 \times 1.5 \times 18 \text{ hours/week} \times 52 \text{ weeks} \times 50 \text{ floors}$$
$$= 2,386,800 \text{ KWH}$$

At $.02772/KWH, this is an additional saving each year of $66,162.

Most building operators bill tenants for after-hours lighting use based on the lease agreement, and most base those billings on a record kept at the security desk on the first floor. If the relays which are installed have an extra pair of contacts, these can be used to return status to the CRT of the EMCS; that is, whether the circuit is "on" or "off." The calculated point program of the computer can then be used to accumulate after-hours use, calculate, and print out a bill for the tenant.

SUMMARY OF ENERGY SAVINGS

The individual savings computations in this section were based on computing savings based on estimated reduced "run time" or "on time" and on reduced outputs. These savings are each listed below, together with the total.

KWH and fuel savings	$78,752–$112,236 per year
Demand Savings	$78,876 per year

These figures can be compared with the 10–15% savings of projected annual energy use discussed at the beginning of this study.

	10%	15%
KWH and fuel savings	$ 66,570	$ 99,855
Demand Savings	58,927	87,941
TOTAL	$125,497	$187,796

It can be seen that the sum of the detailed savings, excluding light-ing, falls generally within the range of 10–15%.

If lighting control is included as a part of the EMCS, additional an-nual savings could range from $33,320 to $49,981.

	Energy and Fuel Savings	Demand Savings
1. HVAC Air-Side		
Start/Stop	$ 8,649–$ 17,298	
Optimum Start/Stop	$ 4,324–$ 8,648	
Night Set-Back and Summer Lock-Out		
Heat	$20,511–$ 41,022	
Fans	$16,355–$ 16,355	
Outside Air		
Supply Air Fans	$ 4,858–$ 4,858	
Exhaust Air Fans	$10,285–$ 10,285	
2. Chiller Control		
Start/Stop	$ 3,847–$ 3,847	
Chilled Water Reset	$ 4,933–$ 4,933	
Cooling Tower	$ 4,990–$ 4,990	
3. Demand Monitoring and Control		
Summer Demand		$66,339
Winter Demand		9,537
TOTALS	$78,752–$112,236	$75,876

Total Electricity Cost Savings (KWH, fuel, demand, tax)
$154,628–$188,112

4. Lighting Control	
one point per floor	$33,320–$49,981

21

Case Study 2:
Energy Management
and Control Systems
in Supermarkets

Jon R. Haviland, *P.E., C.E.M.*
Ralphs Grocery Company *

Ralphs Grocery Company is a progressive chain of about 95 stores operating in Southern California. Ralphs began operation in 1872, and was a family-owned business until 1968 when it became a division of Federated Department Stores. In addition to the stores, Ralphs operates associated manufacturing, warehousing, and office facilities.

SUPERMARKET ENERGY USE

Energy use in a supermarket is quite different from that in other types of commercial buildings. Table 21-1 shows a breakdown of energy in our stores. There will be some variation depending on the design of the store, but the relative percentages shown are typical of all supermarkets. As would be expected, the majority of the energy use, 60%, occurs in the refrigeration system. The second largest area is lighting, with 21%. The heating, ventilating, and air conditioning system accounts for only 8% of our consumption. This is primarily due to the use of heat reclaimed from the refrigeration system to provide

---

* Since writing this chapter, Mr. Haviland has moved to Robinson's, Inc., in Los Angeles.

Table 21-1. Average Store Energy Use

1.	Refrigeration	. . . .	60%
	A.	Compressors	44%
	B.	Condensor	4%
	C.	Case fans and warmers	8%
	D.	Case defrosts	4%
2.	Lighting	. . . .	21%
	A.	Sales area	16%
	B.	Case lights	2%
	C.	Support areas	2%
	D.	Building exterior	1%
3.	Heating, Ventilating, and Air Conditioning	. . . .	8%
4.	Others	. . . .	11%
	A.	Checkstands	
	B.	Meat preparation	
	C.	Office	
	D.	Delicatessen	
	E.	Hot water	

all heat for the store. The remaining 11% is consumed by areas such as the checkstands and registers, meat preparation, delicatessen, etc.

One important point to note from this analysis is that most of the energy is process-oriented, rather than building oriented. Thus, our requirements for temperature control, etc., are stricter than most other buildings where the only potential problem is discomfort. The potential for product loss or spoilage requires a more thorough evaluation and slower implementation of some changes than would be the case in other buildings. This also means that a smaller part of the total energy use is available for control.

ENERGY CONSERVATION EFFORTS

Ralphs first became involved in energy conservation in 1974 during the first Arab Oil Embargo. At this time, the City of Los Angeles enacted statutes which required a reduction in energy consumption, as well as giving some specific limitations on use. We operated all our stores in this manner because all California utilities are heavily dependent on imported oil for power generation. The results of this were impressive: 17% average reduction, with a range of 10% to 34%. This improvement was primarily a result of reduced lighting levels and elimination of decorative lighting, much of which has been reactivated, as well as increased employee awareness.

In 1976, Ralphs realized that additional conservation could only be achieved by making technical changes in the stores. Two areas were identified for further testing and evaluation: computerized energy management and control systems and use of heat recovered from the refrigeration system to provide store heat and hot water. In addition, store design was reviewed and changes were made to increase the energy efficiency of the building.

INITIAL ENERGY MANAGEMENT SYSTEM EXPERIENCE

During early 1976, Ralphs started to consider the feasibility of and potential benefit from installing computerized energy management systems in the stores. A review group was formed which included Ralphs' personnel and consultants from Federated Department Stores, our electrical engineering design firm, and our refrigeration firm. This group reviewed the potential benefit from any type of system, and also reviewed the available hardware.

The final decision came down to a choice between a stand-alone microprocessor, the Monitrol MP-2 manufactured by CSL Industries, and a centralized system using the IBM System 7. The central system offered better control and more information capability, while the stand-alone system had some first cost advantages. The central system also required considerably higher operating costs for maintenance and leased phone lines. One other factor which influenced the final choice was a feeling that very little use would be made of the extra capability of the central system when many stores were on-line.

Thus in late 1976, one system was installed in one of the stores. The only loads which were controlled were the refrigeration compressors. Staff personnel spent a considerable amount of time in the store monitoring case temperatures and system operation. A fairly aggressive cycling strategy was used. An average reduction in KWH consumption of 8% was achieved, although not all possible loads were controlled. One area with a great potential for savings was lighting. This was expected to be fairly easy to implement which was why it was not included at this time.

At the same time, another system, manufactured by Entech, was installed in a store in Northern California. This system was installed, operated, and monitored by the manufacturer. All possible equipment in the store was controlled. The control strategy was generally more aggressive than could reasonably be used when people weren't in the store constantly. This system did achieve 16% reduction in KWH consumption.

PROGRAM IMPLEMENTATION

As a result of the two test installations described earlier, Ralphs decided to proceed with the installation of systems in twenty stores. A 12% reduction in KWH consumption was expected, based on the average of the two tests. Installation was completed in all stores in early 1978, although only the refrigeration and air conditioning compressors were controlled. Lighting control was to be completed at a later date. In addition, the cycling strategy tended to be less aggressive because people could not be in the stores constantly.

Initial savings estimates were based on an historical comparison of billing data. This indicated a reduction in consumption of only 6%. This was probably due to the less aggressive strategy and the failure to immediately implement lighting control. In addition, there were serious reservations about the historical comparison methodology. There are many factors which affect energy use in a supermarket: hours, sales, weather, manager involvement, etc. Most of these change from year to year, and it is difficult to isolate and quantify the effect of each.

The next step we took was to implement lighting control in some of the stores. This is the area which causes the most variation in installation cost. In some of the stores contactors had been installed during construction, and all lights were on a few electrical panels which were centrally located. In other stores, there were lighting circuits on many panels scattered throughout the store. This also led to a considerable variation in savings depending on how tightly the store personnel had previously controlled the lighting. One problem which was encountered was how to provide an even, reduced light level for night stocking periods.

After lighting control was completed in a few stores, we proceeded to perform a controlled test. The methodology adopted was to alternate short periods with all control points in "on" (override) with like periods with all control points in "auto" (controlled). The procedure involved a two-week test with changes made and the electric meter read on Mondays, Wednesdays, and Fridays. Figure 21-1 shows the results of this type of test in one typical store. The advantage of this methodology is that it isolates the effect of the energy management system and minimizes potential variation in the other factors which affect energy consumption. The disadvantage is that it may overstate the savings, depending on the extent of manual control of the lighting in the store.

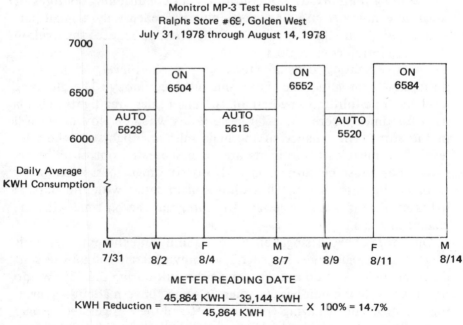

$$\text{KWH Reduction} = \frac{45,864 \text{ KWH} - 39,144 \text{ KWH}}{45,864 \text{ KWH}} \times 100\% = 14.7\%$$

Fig. 21-1. Monitor MP-3 Test Results

The results of these tests demonstrated to top management the soundness of the approach and the excellent return on investment. In late 1978, we were authorized to proceed with installation in 40 additional stores, which was completed in June 1979. In late 1979, installation was authorized in the balance of the Chain, which was completed in February 1980. Tests have been completed in a total of 26 stores. The average reduction in KWH consumption is 13.1%. The payback when we began the second phase was about 1.5 years, but it is now about 0.75 years as a result of increased electric rates. We generally see some reduction in KW demand also which produces some additional savings.

TYPICAL INSTALLATION

A typical installation in one of our existing older stores requires 10 control points (16 control points are available). Five of these are for lighting, four for refrigeration compressors, and one for the main air handling unit. We do not control the air conditioning compressor separately now because air conditioning is needed only a small part of the time, and allowing it to run when needed reduces the load on the refrigeration compressors.

The five lighting loads are stocking level, sales level, display case, meat case spotlights, and signs and outside canopy. The stocking level has one-third to one-half of the main sales area lights. This is used for times before and after store hours when employees are still in the store. The balance of the main sales area lights are the sales level. The meat case spotlights are on a separate control point because they must be turned on early for the meat personnel to go through the case. We install a photocell in series with the outside lights so that it is not necessary to reprogram the on time with the change in season.

Our stores have a single air handling unit, and the entire store is treated as a single zone. In general, we only program the unit to shut down at night, and do not duty cycle the unit during the day. We do not duty cycle the unit because there is some concern about aggravating the problem of having cold aisles around the refrigerated display cases. Some other chains do, however, cycle the unit. In many stores, we have cold aisle problems if we shut off the unit at night, especially during cold weather.

Our refrigeration system is fairly unique, consisting of five 30 or 40 horsepower open compressors. This system was designed to operate continuously, and this made it more reliable than using many small compressors. While this type of compressor is more efficient than the semi-hermetic type generally used, the overall system is not the most energy efficient. Generally, several case systems are served by each compressor. One compressor, serving the ice cream cases, is not connected to the unit. In many stores, the compressor serving the meat cases is programmed not to cycle because problems were encountered with product quality.

Table 21-2 shows the load programming for a typical day. The store hours are 8:00 a.m. to 10:00 p.m. The stocking time may be different for each day of the week, and Sunday hours may be different. In addition, there is the capability of having a second program. This could be used, for instance, to provide longer cycle off times at night. In general, we adopt a more conservative strategy which generally allows us to program a system and then simply let it operate. This does not provide maximum savings, but it does minimize the time which must be spent with each unit.

The cycling strategy provides a minimum on time that the load will operate between cycles and minimum and maximum off times. The maximum off time is used to dampen peak demand during periods when the system projects that demand will exceed the set point. If available, loads could be programmed with a zero minimum on time which would shut the load down only during high demand periods which would provide more demand limiting. Demand limiting is not a serious concern for supermarkets because the high load factor means that the peak demand is not high above the demand in adjoining periods. In addition, there are very few deferrable loads when the compressors are already being duty cycled.

FUTURE EXTENSIONS

The Monitrol Unit has additional capability which we are not presently using, but expect to implement in the future. First is a defrost

Table 21-2. Load Description and Program

Load	Description	Start Up	Shut Down	Cycle Non-Cycle	Min On	Min Off	Max Off
1-1	Stocking Level Lights	0400	2400	NC	--	--	--
1-2	Sales Level Lights	0755	2230	NC	--	--	--
1-3	Case Lights	0800	2200	NC	--	--	--
1-4	Meat Case Lights	0700	2200	NC	--	--	--
1-5	Outside Lights	1500	2230	NC	--	--	--
1-6							
1-7							
1-8							
2-1	Compressor 1 (Meat Cases, Produce Case, Walk-In Boxes	0000	2400	C	26	6	8
2-2	Compressor 2 (Dairy, Deli, and Beverage Cases)	0000	2400	C	23	6	8
2-3	Compressor 3 (Tub Frozen Food Cases and Freezer)	0000	2400	C	26	6	8
2-4	Compressor 5 (Multi-Deck Frozen Food Cases	0000	2400	C	26	4	6
2-5	Air Handling Unit	0500	2300	NC	--	--	--

interlock capability. This allows the compressor to be forced on or off depending on the type of defrost. It also allows for a recovery period after defrost terminates during which cycling does not occur. It is expected that we could use longer off times to compensate for the periods that cycling is prevented.

The system also has the capability to use an input (temperature, humidity, etc.) to shift from Schedule 1 to Schedule 2. This could be used to reduce cycling based on a rise in temperature or humidity. Thermostats or pressure sensors could be used to indicate high temperatures in the cases, then the program would be switched to a less aggressive cycling strategy.

The most promising capability which is presently under development and testing is two-way communication capability. This would use dial up equipment so that the high cost of leased phone lines is not a factor. Thus, units could be programmed and the condition checked from a remote location. This would eliminate much of the traveling that is now required to check the systems.

It could also lead to more savings because a more aggressive strategy could be used during winter since it would be easy to reprogram all the units as necessary. The major problem with stand-alone systems is that loads may be overridden when it is not necessary, and there is no way to check this without visiting the store. It is also necessary to visit the store in order to reprogram the unit, although some store personnel do reprogram lighting when necessary.

KEYS TO A SUCCESSFUL PROGRAM

The first key to a successful program is to know your equipment, and what can be done with it. This means that the refrigeration service people, either outside contractor or in-house, must be fully involved from the beginning. A poorly planned program can result in considerable product loss and equipment damage. We experienced a few minor product losses due to unforseen problems. Failure of a few old compressors was experienced, but it is generally conceded that these would have failed in the near future anyway. Do not rely on the system vendor to set up control systems or programs.

The second key is vendor selection. One important point to

remember is that all systems will generally provide similar savings. You need to look for a system you are comfortable working with, and has the capabilities and features that you deem important. The reliability of the vendor is an important consideration, as is the installation and service support which is available. Many problems can be avoided if a turnkey approach is used. This places all responsibility in the hands of one company, and keeps you from being in the middle of arguments over who is responsible in the event of problems or damage occurring during installation. This can also minimize the involvement of your staff during installation of the systems.

The third key is continued monitoring and follow-up store visits. In some cases, there is a tendency to override the system whenever the least minor problem is encountered. In these cases it is necessary to build confidence in the system. In addition, some revision to compressor programming may be required depending on actual store conditions. It is also necessary to keep in touch with the refrigeration service people so you are aware of problems they are encountering.

SUMMARY

Ralphs Grocery Company has achieved significant reductions in energy consumption by use of computerized energy management and control systems. The program took several years to develop and prove the concept, and then to completely implement throughout the Chain. Average results are a 13.1% reduction in KWH consumption, providing a payback in less than one year at current electric rates. A relatively simple installation program has been used to minimize the need for ongoing site visits and reprogramming. Additional capability is available which could be implemented, and could provide additional savings.

22

Case Study 3:
Energy Management
and Control Systems
at Princeton University

Michael E. McKay,
Hugh W. Smith,
Princeton University

Princeton University has grown through the years to the point where it now includes two campus locations; a main campus and the Forrestal campus located approximately five miles away. The University has six million square feet of building floor space in 186 buildings ranging in character from single family homes converted to dormitories up to sophisticated genetic research laboratories. A large number of the laboratory and academic buildings were constructed in the early 1900s and have since been modified considerably including the addition of air conditioning and laboratory hoods.

Energy costs for the University as a whole have risen from less than $3 million in FY73 to in excess of $9 million for FY82. This threefold increase is due primarily to tremendous increases in energy prices but also partly to a vigorous campus expansion program that has seen the addition of four new buildings over the past three years with several more scheduled to be constructed in the future.

As energy costs became a larger and larger percentage of the University's operating budget, the administration turned considerable attention to means to reduce these costs. The program that emerged consisted of three levels of actions. The first level included such items as establishing a winter heating policy of 65F and a summer cooling policy of 78F, the installation of time clocks on mechanical systems, lamp removal, and generally improved preventive maintenance.

Level two included changes to our utility distribution systems, academic and administrative buildings. Examples are the installation of a secondary pumping system in campus buildings connected to our chilled water plant, adding attic insulation, the installation of variable volume on some air handling systems, and some additional utility metering.

Level three consisted of more capital intensive projects in the utility area, including the installation of a heat exchanger in the chilled water plant to provide "free cooling" in the wintertime and the installation of an energy management and control system.

The first step in the long road to the installation of the energy management and control system was to perform a feasibility study. A major New York consulting firm (Flack & Kurtz) was retained for this purpose and evaluated all campus buildings to determine the potential savings associated with an energy management system. They then estimated the cost of the system and indicated, using present value and life-cycle costing, that the investment was justified. At that time, which was 1976, the cost of fuel oil was $14.00 per barrel and was expected to escalate at approximately 10% per year which would indicate an expected cost today of approximately $25.00 per barrel. Our current cost for fuel is actually in the range of $35.00 per barrel.

Following this step, the consultants prepared a specification which included three phases of construction. The first phase involved the 20 most energy intensive buildings, and the second phase included the next 20 most energy intensive buildings. All of the buildings in these two phases with the exception of one were academic, administrative or laboratory buildings.

The third phase included some of the less energy intensive administrative buildings and the dormitories on campus. Bids were solicited from the major vendors of energy management and control systems at that time, including Honeywell, Johnson, Powers, RobertShaw and one lesser known company, Hamilton Test Systems. All of the proposals were evaluated on a life-cycle basis which included such items as first cost, annual maintenance and operating costs, and the ability of the proposed system to meet the specifications.

When evaluating the bids for system performance, we used a point system. Each of the proposals was assigned a certain point value under six different categories. These categories included the vendor's experience with similar systems, his willingness to accept the turn-key role in the project, his capability to train our personnel, and the terms of the warranty and service contract.

The next category was basically a system overview which covered the reliability of the entire system and redundancy. The third category was central station performance in which we looked at such things as computer equipment hardware, information displays, the ease with which commands could be given, difficulty in making software changes—basically, the human engineering aspect of it from the system operator's point of view.

The fourth category concerned energy conservation software in which we evaluated the bids based on the software's capability to perform start/stop, enthalpy control, cycle, load-shed, hot and cold deck reset, and other optimization functions.

The fifth category had to do with field hardware where our concern centered on noise immunity, ease with which troubleshooting could be performed, and fail safe operation.

The last category covered special requirements such as the ability of the system to handle fire and security information which we eventually deleted.

After this evaluation, Hamilton Test Systems was selected as the contractor for the job and a contract was signed in August 1977. Phase I construction began at that time and approximately six months later we authorized the beginning of Phase II construction. As the time approached to make the decision on Phase III, we reevaluated the potential savings and found that the ability of an EMCS to save energy in our dormitories was extremely limited and subsequently Phase III was dropped. Only one dormitory which included dining halls, common areas, and which is air-conditioned, was included in the second phase. The entire system installation was completed in April 1980 and we have been operating successfully since that time.

The energy management and control system called CSCS, or Central Supervisory Control System, consists of a central computer located in our Physical Plant offices that is linked primarily by direct burial tri-axial cable to 38 buildings, and via modems and leased telephone lines to two other buildings. Within each building is located a RMDM or remote microprocessor data multiplexer.

These panels communicate with the central computer and perform a limited control function within the building. All wiring from the RMDM to the individual points, i.e. motor starters, temperature sensors, valve controllers, is with twisted pair or in some cases twisted shielded pair for sensors and analog signals.

In a typical mechanical system that is controlled by the CSCS, the supply fan is directly controlled by a normally closed relay to the automatic leg of the hand/off/auto switch. The return fan is subsequently interlocked with the supply fan. Temperatures in the space and duct work are sensed using RTD type sensors and the signals transmitted back to the RMDM.

Closed loop calculations are performed in the RMDM which then transmits signals to all of the analog control points, including the damper motors and heating and cooling coil valves through a pneumatic control panel. A pneumatic control panel contains three-way solenoid valves and transducers to convert the 4–20 milliamp analog signals from the RMDM to 0–18 PSI pneumatic signals going to the valves. It should be mentioned that all of our analog controls on air handling and water systems are pneumatic controls.

All communications between the buildings and the central computer are transmitted on direct tri-axial cable or on leased telephone lines. The RMDM also serves as the communication interface in this link. Information accumulated by the RMDMs such as temperature readings, status of mechanical equipment, steam, chilled water, and electrical meter readings are collected in the RMDM and signals are then multiplexed and transmitted back to the central computer.

In some cases, such as with steam, chilled water, and electricity consumption, the information is stored on discs later to be used in reports. Other information such as temperatures and fan status, is available to the operator via the CRT, and is used by the computer in open-loop calculations.

Closed-loop calculations are typically those performed within the RMDM located in the buildings and are used to reset valves, damper motors and inlet vanes to maintain a predetermined reference duct or water temperature or static pressure.

Open-loop calculations, on the other hand, are performed within the computer itself in our Physical Plant building and establish this predetermined reference. In a typical open-loop calculation, the computer would compare the space or room temperature within a portion of the building to the heating or cooling reference for that building and then calculate the appropriate reference temperature to be maintained on the fan discharge.

This information is then transmitted back to the RMDM and resides there for a short period of time (15 minutes) and is used in the closed-loop calculation until the next time the open-loop calculation is performed. Certain control functions are reserved exclusively for the computer, as opposed to the RMDM, including starting and stopping of mechanical equipment, time of day scheduling, and optimal start time calculations. The RMDM has a very limited control capability and cannot stand alone for more than 15–30 minutes operating effectively without communication to the central computer.

At any point in time, the operator can interrupt the control loops and take manual control, using the CRTs that are connected to the central computer. He has the ability to start or stop mechanical equipment, open or close valves and dampers, and to interrupt the control logic and establish a reference temperature for any of the closed loops. These options become extremely important in troubleshooting mechanical systems to determine whether or not the problem is in the computer, any of the control operating devices, or the mechanical system itself.

This can save tremendous amounts of time when dispatching mechanics to repair problems. It also becomes important when making software changes to optimize performance of some of the mechanical systems because the operator can simulate what would happen if he changed some of these parameters in the data base and make sure that the system will operate properly afterwards.

Each of the pneumatic control panels contain three-way solenoid valves in addition to the transducers. These three-way solenoid valves play a very important role in the architecture of our system. When energized, or enabled, which is the normal condition, these solenoids transmit the pneumatic signal from the CSCS transducer to the pneumatic device, such as the valve or the damper motor.

When de-energized, or disabled, the signal that passes to the valve or damper motor comes from the local pneumatic control panel. For the most part, our system is a retrofit system and was added on top of existing controls that have been in use in the building for years. A single signal from the RMDM either enables (energizes) all of the solenoids on a system or disables them.

Again this becomes important in certain situations. In the event one of the CSCS sensors should fail to operate properly, the operator can de-energize these solenoids and turn control back to the building controls. Of course, this means that we lose our energy management functions, but the building can continue to function comfortably without immediately dispatching a mechanic to solve the problem.

Should the communication link between the RMDM and the central computer be broken, we again automatically de-energize these solenoids. In the unlikely event that we had no computer to communicate with the RMDMs, all of the solenoids on campus would become de-energized and the control would revert to original building controls. This provides a great deal of flexibility in that we have a backup control system for all of our computer controls.

All of the communication links are redundant. We have two cables to each RMDM from the central computer. If one should fail the other one automatically takes over and continues the communications. Within the central computer room, we have redundant computers. If for some reason there should be a problem with the main computer, the backup computer will automatically take over until the problem with the main computer can be resolved.

It is relatively easy to compare the normal pre-energy-management operation for a typical system with the operation after the CSCS is applied to the system. In a typical dual duct system, of which we have a large number, a hot deck and a cold deck temperature are maintained and supply air to each zone is reset by the local zone thermostat and a mixing box. In many of these systems, the cold and hot deck are operated at 55F and 90–110F respectively. This is done to accommodate the range of temperature requirements that might be experienced by the different zones.

At all times, the system is providing cooling (either with chilled water or outside air) and heating using steam. With the CSCS, typical zone temperatures are sensed in the hottest and coldest zones in the building. These temperatures are then used to determine whether or not any hot deck or cold deck is required and only the minimum amount required is provided. On a typical 30° day the hot deck on a system without energy management would be 100F, the cold deck would be 55F. Using the energy management system, the hot deck would be maintained at approximately 70–80F and the cold deck would be 68F using return air only. Considerably less energy is being consumed in the latter case.

Another typical system is one in which we have an air handler with cooling and terminal reheat. This system is one that is located in one of our office buildings and when it was installed it was designed to operate with a 55F fan discharge year round and 200F water to the reheat coils, provided by a steam to hot water converter. Even on a 65–70F day it is likely that this system would be both cooling at the central fan and reheating at the terminal reheats.

After application of the energy management system, cooling would only be provided when it was found that one of the zone thermostats connected to the computer was calling for cooling and at that point it is extremely unlikely that any of the zones would be calling for heating, so no hot water would be provided to the reheats. Generally speaking, in a system like this, only the cooling or the heating is in operation at any one time and for a good portion of the year neither one is provided and we operate strictly on return or outside air.

There are, as noted above, a number of energy management functions performed by the computer itself. One of these would be optimal start. The optimal start calculation in the computer is performed beginning at a time several hours before the building is to become occupied. Space temperatures in the building are sampled and compared to the desired temperatures in the upcoming occupied period.

Using this information as well as the outside air temperature and certain factors pertaining to the mechanical system and building, the computer determines the optimal time to start the system in order to provide the desired temperature by the time the occupants arrive for work. The mechanical system is then started up earlier if required, but only if required; whereas without the CSCS we would typically have time clocks set for two or three hours before the people come into the building.

Occupancy schedules also become important in our facility. Prior to the CSCS, the most sophisticated kind of control we had for occupancy consisted of time clocks. Even with the use of seven day time clocks, most systems were started at 6 or 7 in the morning to ensure that they were heated or cooled to the proper temperatures by 8 or 9 and left on until 5 or midnight depending upon the type of building.

The only way to modify these schedules was to send a mechanic out to the building to change the time clock. Using the CRT in the central control room, the operator can, in a few seconds time, change the occupancy schedule for any system for any day and we can thereby schedule the use of our buildings on a daily and hourly basis. A good example of this would be a lecture room that is used only for special events. The operator feeds the schedule into the computer on a weekly basis and can run the system for the two hours required for a lecture or for an entire day if necessary but does have the capability of only running the system when it is actually required.

Load shedding, of course, is an option that is available with the central computer system. The computer receives a signal from our main substation indicating the instantaneous demand level we are drawing from the utility. Since we are on a 15 minute demand interval, the computer continuously calculates the projected demand for the end of each 15 minute demand period.

If this calculation indicates that we will exceed a predetermined demand level, the computer then sends out load shed signals. This load shed signal will, in some cases, turn off mechanical equipment that is operating and in other cases, will extend the off time if a piece of mechanical equipment is already in cycle.

Cycle, of course, is a brief period every hour when we normally turn off the mechanical equipment, generally, 15 minutes to a half-hour out of every hour.

Depending upon the type of facility, load shedding and reduction of demand charges can be significant. In our case, we feel that it has not been as beneficial as hoped because the mechanical equipment is a fairly small percentage of our load and because the cycle program will normally have 25% of our mechanical equipment off at any one time anyway.

Another function referred to previously was the use of the computer to accumulate consumption data in the buildings. With this option, we can provide hour by hour read-out of the steam, chilled water and electricity consumption for each of the buildings on the system. This helps us to determine where, for instance, a large part of our electric demand is going. These reports have been used extensively as an energy management tool.

One of the significant benefits of this system is the ability that it gives to one person to check up to 40 buildings by simply sitting in a chair and pressing a few buttons. We have found a large number of previously unidentified problem areas and sources of energy waste by simply looking at these buildings during unoccupied periods or even occupied periods.

A good example of this is steam heating coils. A decision was made early in the design of the system not to include steam preheat coils on computer control because they are used to prevent freeze-ups and maintain a constant 45F leaving air temperature. Therefore, little optimization is possible. Normally, any rise in temperature above that point is done with the heating coil that follows the steam preheat coil.

In checking the buildings, it was found that most of these steam preheat coils have normally open valves that would open whenever the fan shuts down. After identifying these cases, we were able to take corrective action by automatically shutting these coils down when the outside air temperature exceeded 40F.

No energy management system is without problems, and we have our share of these. The problems stem primarily from the fact that we are trying to operate mechanical systems designed in the days of energy inefficiency in an energy efficient manner.

One of the problems that has cropped up is the fact that in many of our mechanical systems we supplied too much air to the occupied spaces. Prior to the energy management system this was not a problem because the supply air would be cooled or heated sufficiently that it did not cause any discomfort. However, after application of the energy management system, we experienced problems during the heating season in that the temperature of the supply air was considerably reduced, creating a situation wherein the occupants of the building felt drafts.

Solutions to this problem ranged all the way from application of variable volume systems and reducing the volume of air during the heating period, to extended cycling of the fans so that when the fan system did run it was supplying a higher temperature air for a shorter period of time and the same total amount of heat was supplied to the space. In a number of our buildings, we have fan coil systems. These were designed so that when the room thermostat called for heat, a valve would open and hot water would flow to the fan coil unit while the fan motor was blowing. Our energy management system stops the pumps when heat is not required in the building.

The result is that the fan coil units continue to blow air but at reduced temperatures of approximately 65–68F (recirculated air) and again create a chill or draft problem. The solution to this problem involved tying all of the fan coil unit blower circuits into the same contactor we used to stop the pumps so that the fan coil blower stopped with the pump.

Steam systems do not lend themselves well to control optimization but we did find that the measured savings indicated considerable success. While there is no optimization that can be accomplished on a single valve steam radiation system, it is much easier to maintain a lower level of temperature in the building when centrally controlled than when the building is under control of a weatherstat or other such system.

One of the ways we averted a number of problems with our system was to insist on a point to point check. This consisted of verifying the operation of, or reading from, each point in the field and comparing it to what the operator read on the CRT.

This process took several months to accomplish, but the results were well worth the effort in that we found a large number of sensors that were out of calibration, a large number of fan status relays that were not providing true status, and a certain number of analog control signals that were not being properly transmitted through to the valve or damper. It has also served as a valuable tool for our maintenance people in that they were able to see where every single point was and come to a better understanding of how the system operated.

Our single biggest problem with this system was not inherent in the system itself but in our application of it. During the construction of the energy management system, we also had several new buildings under construction and specified that these should be tied into our central system at the same time. As mentioned before, our system consists of 40 buildings with 100% backup for every component and one of these backup modes consisted of the local controls backing up the CSCS controls.

As we added two more new buildings to the system, we felt that it was only practical to require local pneumatic controls in these buildings as a backup to the computer to be consistent with the rest of the system. We were wasting a considerable amount of money to purchase controls that would rarely, if ever, be used. We felt, however, that in the event of a computer failure, all of the buildings on our system should be able to operate independently.

This led us to make some inquiries of Hamilton Test Systems and many of the other vendors from whom we had received proposals as to what could be accomplished in the way of energy management on a smaller scale, independent of our present system. We found that most of the vendors were providing distributed processing at this point and after a careful search selected Hamilton Test Systems again for our next generation of energy management.

This next generation of equipment, the unit controller, is a much different concept from our original system, primarily in the system architecture. The unit controllers are truly distributed processing in that most if not all, of the actual control functions, calculation, and scheduling, are performed in the building panel which is analagous to the RMDM. In fact, the unit controller can and does operate completely independently of a central computer but can eventually be tied into a central computer.

We have found a number of significant advantages in the newest generation of energy management systems. Installation costs are reduced for a number of reasons. The unit controller is in itself considerably less expensive than the RMDM. Connection to the central computer, if one is installed, is also considerably less expensive because there are lower cabling costs; for the most part communication is via telephone lines.

There is increased system modularity, which is a major consideration when looking at system reliability. In our present energy management system, the CSCS, the loss of two major communication lines to any one set of buildings or the loss of both computers results in the loss of any effective control function performed in those buildings. However, with the unit controller system, we can lose the central computer or telephone communications to it and it will still stand alone and operate with 99% effectiveness.

Because of our reduced exposure to failure with this system's architecture, we have elected not to install any backup controls in new construction buildings and we are in a position where we no longer feel that it is necessary to have redundant communications between the buildings and the central computer; nor are we required to have redundant central computers.

Present plans are to use the unit controller as a basic building block in expanding our future energy management capabilities. They will be installed in two types of buildings, one will be retrofit in existing buildings where the existing control system is either extremely wasteful from an energy point of view or in need of replacement from a maintenance point of view. The other application will be new construction projects where the unit controller will be applied as the only control system in the building.

At this point in time, we do not have a significant need for a large number of these but expect this to develop over the next few years because of our renovation and construction program. Rather than wait until we have 25-30 buildings to put on unit controllers and purchase a central computer along with the field units at one time, we can begin realizing the energy savings in some of our existing buildings by installing unit controllers individually.

At some point in the future, when we feel that a central computer is warranted, we can purchase the computer and obtain the additional benefits, but in the interim we will be achieving energy savings. The unit controller also allows us to perform energy management functions in smaller facilities that would not warrant a large RMDM.

We look at the unit controller as a basic building block. It has stand-alone capabilities and all of the unit controllers are interchangeable because they have the same basic programming and ROM (read only memory). The data base for each particular application is loaded, in our case, by the owner's module. This is important when evaluating different energy management systems. Many systems claim to be stand alone and can operate without a central computer, which is to some extent true.

However, many of these require down-line loading from a central computer that you either purchase or hook into by telephone lines at the vendor's factory. The unit controller, on the other hand, has the ability to be loaded individually and we are not relying on anyone else or a central computer to keep a building on line. At least at this point our vendor has indicated that three types of unit controllers will be available, varying in the number of points each is capable of handling.

Some of the other less significant but extremely convenient features of the unit controllers are that the optimization program runs more frequently than in the central computer, providing for closer adherence to scheduled temperatures. In many large computer installations, temperature sensors in the field must be calibrated by the installation of resistors on the field hardware and this becomes an extremely time consuming and inaccurate way to calibrate them. In our unit controller, we can recalibrate temperature sensors simply by punching into the unit controller the actual temperature measured in the field and the bias is adjusted automatically.

The unit controller is a new development for Hamilton Test Systems and like most universities we are fairly conservative in our approach to acquiring newly developed equipment so we purchased and installed one in the Physical Plant building to watch the operation and determine whether or not it truly met our needs.

This unit controller has been in operation for approximately six months and we found that it does an excellent job. At this point, we have construction documents out for a new dining hall to be constructed on campus and plan to use the unit controller as the only control system for this building. As much as we had confidence in Hamilton, based on previous dealings with them and our experience with the unit controller, it was additionally reassuring to know that Walt Disney World, Inc. had selected the unit controller as the sole control system for EPCOT (Experimental Prototype Community of Tomorrow).

RESULTS

While we have just begun application of the unit controller on campus and have no measurable results, we have spent a considerable amount of time analyzing the results of our several years of operation with the central supervisory control system. Like most institutions that have a large multi-building campus, we did not have the luxury of metering all of our utilities prior to the installation of the energy management system.

This makes it extremely difficult to determine whether or not the anticipated savings were achieved. Considerable time and effort was spent on our part recently to determine whether or not the anticipated savings were achieved, and we did this by two routes: the first was an in-house analysis followed by an independent consultant's analysis of our utility plant reports. We found some extremely interesting information.

The first area that we looked at was electrical because, fortunately, each individual building has its own electric meter. We were surprised to find out that it was almost impossible to determine whether or not the savings were being achieved. For each of the 40 buildings on the system, we analyzed the electrical consumption pattern for the seven years prior to the application of the energy management system and determined the mean and standard deviation of the total annual kilowatt-hours of consumption.

In almost every case, we found that our calculated savings were smaller than the standard deviation and therefore determined that we could not reach any decisive conclusions. Table 22-1 shows some of the data we obtained from the mean and standard deviation calculated based on fiscal years 72 through 78 (the before data) and compared to the actual FY79 and FY80 consumption. The last column is the original savings figure calculated by the consultants for each of the buildings shown.

Table 22-1. Annual Electrical Consumption by Building
Before and After the CSCS

| Building | FY72-78 | | FY79 | FY80 | Estimated Savings |
	$\bar{X}$	σ			
3	603.2	127.0	411.2	352.1	199.5
46	739.5	116.6	532.3	499.8	175.7
62	1,427.0	143.1	1,461.2	1,459.6	122.0
84	1,131.4	423.0	1,306.3	906.6	127.8
98	509.4	175.7	378.0	198.0	323.6
Totals	4,410.5	- -	4,089.0	3,416.1	948.6
(All data above is in thousands of KWH)					

On a building-by-building basis, it is hard to determine whether or not any savings were actually achieved given the statistical significance of the data available, but on the whole it appears that our consumption has been reduced far more than expected. Of course, there are other energy conservation projects such as relamping, voltage reduction, and rewiring of buildings that have occurred during this period that have also contributed to the savings.

The next area examined was our cooling load. In this case, we looked primarily at our summer cooling. While we do run our chilled water plant year round, our summer load is the only one affected by an energy management system. The rest of our load is a continuous year-round load for cooling of experimental equipment and computers.

Once again, detailed data is not available. However, using the central plant data, we compared our total campus summer-time cooling requirement before and after the CSCS. The before period consisted of the five years from 1974–78, and the after period consisted of the single year 1980. The data was plotted on a graph where total monthly ton hours of cooling were compared to cooling degree days and we found a significant decrease in FY80 as shown in Figure 22-1.

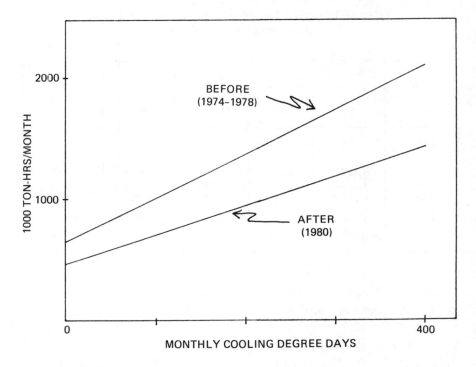

Fig. 22-1. Comparison of Cooling Requirements Before and After CSCS

A similar analysis was performed on our steam consumption. If anything, the metering problem was amplified when it came to the steam savings. Whereas most of the buildings that were connected to our central chilled water system were also on the energy management system, only 50% of our connected steam heating load was operated by the energy management system.

In the case of chilled water, a 10–20% reduction in chilled water consumption in each building resulted in a 10–20% reduction in our total chilled water usage. In the case of steam, however, the 10–20% reduction in the use of steam for heating purposes in 50% of our connected load would only result in a 5–10% decrease in steam usage produced by the boiler house.

By performing an analysis similar to that for chilled water, we found, however, that the expected savings in steam for heating purposes were being achieved. For each of the main steam lines that are separately metered, a comparison was made of before (FY77–79) and after (FY81) data. The data used was heating degree days per month plotted against 1000's of lbs of steam per month.

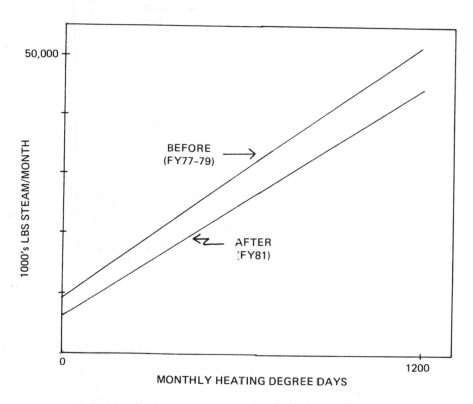

Fig. 22-2. Comparison of Low Pressure Steam Requirements
Before and After the CSCS

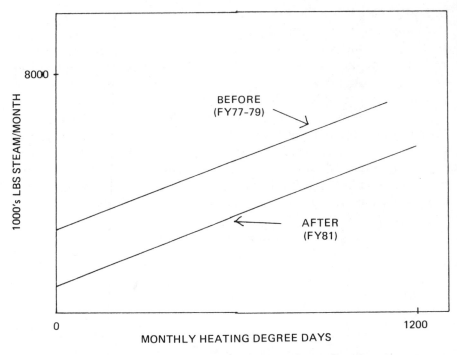

**Fig. 22-3. Comparison of Medium Pressure Steam Requirements
Before and After CSCS**

Figures 22-2 and 22-3 show the least-squares fit to the before and after data for two different steam lines. In Figure 22-2, the data is for a low-pressure line. Only a small percentage (25%) of the buildings on this line are controlled by the CSCS. Figure 22-3 is for a medium-pressure line on which 50% of the buildings are CSCS-controlled.

A significant factor when comparing the before and after data for an energy management system is the ability to isolate the effects of the energy management system as opposed to the effects of all the energy conservation measures taken as a whole. As mentioned earlier, we have had an on-going energy conservation program in many different areas beginning in 1973. Other attempts were made to isolate these effects using a Btu balance for our total plant in which we estimated the savings from each of the projects we accomplished and compared those savings to our annual Btu consumption before and after the project. While this is a somewhat crude approach and relies heavily on the accuracy of our estimated savings, we found we were achieving the anticipated reduction for each project, including the CSCS, within 10–15% of the calculated figure.

The total estimated savings from our system were 825,000 gallons of oil and 4,500,000 kilowatt-hours per year of energy. In terms of current energy prices, this represents an annual savings of approximately $1,000,000. The initial cost of our system was slightly over $3,000,000. Taking into account operating costs we feel that we will have achieved approximately a four-year payback on the investment.

23

Case Study 4:
Automation . . .
Shedding a Little Light
on Building Security

This book has illustrated that the face of computerized building control is in a state of continual change. Just when we thought that automation had energy management as its sole function, market conditions, combined with the efforts of bright engineers and astute building owners, have resulted in a variety of new applications. As building owners develop a greater awareness of the capabilities inherent in their automation systems, new ideas abound: information management, intelligent buildings/system integration, office automation, tenant off-hour billing and a host of other applications.

This chapter presents a generic case study on one of the more interesting innovations to grow out of this trend in building computerization: Automation/Security Interface. In one retail store chain, users have begun to integrate these building systems, thus deriving benefits that neither system can provide by itself. This application is exciting because it enhances the effectiveness of the store security, while showing tangible evidence that building automation can produce benefits above and beyond energy savings.

The gist of the application is quite simple: the automation system monitors two security points 24 hours a day, one of which signals a breach of security. Upon sensing a breach during the unoccupied mode, the automation system turns on all the lights. This provides a crime deterrent which is considered quite effective by the company's loss prevention personnel. What makes this an exemplary application, however, is the judicious blend of security and energy management features which are applied through effective use of facility automation. The following outlines the range of features implemented in this building, as a case study in ingenuity through state-of-the-art building control.

AUTOMATION OVERVIEW

The market climate in the late eighties is the setting for this case study. It is important to briefly discuss the automation industry at this time. Computerized building control began to enjoy widespread application as a result of the energy shocks during the seventies. Automation customers saw energy costs comprising a significantly larger portion of their operating budgets. Under these crisis conditions, many technologies flooded the market, including computerized controls: "Black Boxes" or "Energy Management Systems."

These systems had greatly varying degrees of success regarding their ability to save money. However, there are other "new" types of benefits that these systems can produce. It is important to consider these "new benefits," because system buying decisions are no longer made based upon a crisis mentality about energy consumption.

Further, property owners are faced with myriad technical and logistical concerns, regarding the performance of automation programs. As a result it is necessary to evaluate automation programs, and answer such questions as: Have the systems installed achieved their full potential, and are there benefits provided by the system which were not originally considered? The answers to these questions contribute to an answer for the ultimate query: Do automation systems have a place in the buildings of the future?

AUTOMATION TECHNOLOGY

A technology which was spawned under crisis conditions has grown to provide building control above and beyond its original conception. The ultimate in facility control has been discussed widely as the intelligent building. While it is true that technical considerations and cost constraints have resulted in mixed reaction and limited applications of the intelligent buildings concept, it remains a viable approach to automation.

More importantly, a need for global building functions still exists in spite of the delays associated with acquiring a workable intelligent building. This is the answer to the questions of a future for automation systems.

The 1980s are touted as the Information Decade, and property owners realize that a facility computer system, or an integrated group of computers, is their greatest ally in the quest for useful building information. This is the approach that today's managers, with needs for global building functions, are taking to solve their information and control problems. They are combining the abilities of discrete building systems to achieve a hybrid intelligent system or what has been called an "integrated system." There are many examples of this type of ingenuity. The integration of security and automation systems outlined here is among the most interesting.

AUTOMATION/SECURITY APPLICATION

As automation systems develop greater levels of confidence in a given company, it is common for departments, outside the energy group, to suggest expanded uses for the system. This was the case with a Fortune 500 Retailer which implemented its automation/-security program. As mentioned. the security aspect of the system energizes the lighting on a breach of security during the unoccupied mode. In so doing, the security professionals in this company plan to give many a would-be thief, in the midst of building penetration, serious cause for concern.

The application goes beyond this, as the security system also monitors the opening and closing of the building's outside doors. The energy group gave this some thought, and decided to pick up a second input from the security system to indicate whether the doors were open or closed. In this way it is possible to sequence lighting and HVAC control from the actual moment of arrival, rather than on a fixed time schedule. In this way the addition of a feature which could save thousands of dollars each year in dollars lost to theft also provides concrete savings.

System Equipment

In the case of both the security and automation systems, state-of-the-art equipment was used. EMCS professionals were called upon to design the application, and to implement the interface. A concerted effort was applied by the manufacturers and dealers providing the systems, as well as in-house security and energy management staff.

The automation professionals stipulated that the interface must utilize intelligent devices. This was considered essential to both system function and reliability. Therefore, points were interfaced to smart field panels using direct digital control technology to monitor and control associated loads. The physical interface between the two systems is effected in an isolated panel. This limits the opportunity for installation problems, as the technicians can terminate wiring in an area separate from the critical operations of each system. Automation and security devices are checked for performance at start up along with all other points, and interface conditions are simulated to verify the operation.

Interface and Function

Though emphasis was placed on intelligence, the interface architecture was quite simple. As mentioned, two points of information from the security system are monitored by the automation system: Door Contact and Security Breach.

Through the security breach point, the automation system monitors the alarm condition around the clock. During the occupied mode, this condition is logged but no action is taken. An option during occupied or unoccupied modes would be to provide a secondary alarm call-out.

During the unoccupied mode, a signal from this point causes the automation system to energize all interior lighting. Lighting will remain on for a pre-set period or until the alarm condition is cleared. Crime deterence has been noted as a primary benefit to this feature, but an additional safety issue is involved. Prior to this interface, the store manager and the police would arrive at a building in darkness, in response to a security alarm. Now the building is well lit; it is possible to determine whether anyone is still in the store, and thus avoid entering a darkened building under potentially dangerous conditions.

With the second point, the building security system monitors entrance and egress via a contact switch at the main door. Initially this was to be a magnetic contact door switch, but in this particular case the company decided to install roll-up gates in front of the door. Therefore, the security system monitored a dry contact which indicated whether the gate was up or down. Upon arriving in the morning a store manager raises the gate, signaling to the security system and in turn the automation system that the day has begun. This event is monitored by the automation system instantaneously, allowing lighting and HVAC equipment to enter the occupied mode.

There were several applications issues to be considered and addressed in effecting this control. The lighting in this store is sequenced to reduce the footcandle levels and eliminate display lighting during "employee only" hours. Therefore it was necessary to delete time schedules for employee lights and program them for operation only with the security interface.

Other lighting control was simply modified. For example, the portion of lighting which is on during "customer hours only" had to be programmed with a time schedule, but this schedule was qualified by the security point. So, the lights would come on at the pre-set time provided the security point was energized.

This strategy was also programmed for all similar loads. The purpose of this strategy was to anticipate a condition where a blizzard, or some other phenomenon, kept the manager and store personnel at home. In this scenario the building would remain in the unoccupied mode, thus saving the energy dollars which would otherwise be spent on lighting, heating or cooling.

Another applications issue involves setting the security alarm. It is necessary for the manager to lower the gate, thus de-energizing the security point, to set the alarm. To address this concern, the automation system had to be programmed with a time feature that allows a 10 minute delay prior to shutting off the lights. This feature will "wink" the lights when the security point is de-energized, to signal the start of the 10 minute period.

Start Up and Demonstration

Due to the critical nature of the automation/security interface, a thorough system acceptance must be performed. As noted above, the operation of the system interface must be simulated in occupied and unoccupied modes. This of course requires coordination, as the security system is programmed to call out to the local police station on an alarm. A start-up meeting was therefore held with representatives of each major participant to orchestrate a full-scale simulation. On-going follow-up was also provided to ensure that the combined benefit of the two systems met with the original plan.

Options for the Future

This type of hybrid system offers numerous options beyond those described in this chapter. During the design and implementation process several interesting ideas were considered.

Generating statistical data from a door count was a particularly interesting consideration. The system would use the magnetic door switches for a count of door openings during customer hours. A statistical design could be developed to totalize the data, and generate reports. Such reports would quantify customer traffic through the store, and be useful in a variety of ways. Example of potentially valuable reports are: customer traffic as a function of advertising and sales, peak traffic and sales reports by time of day for use in scheduling personnel, and comparison reports on sales volume and traffic for groups of stores to evaluate specific locations.

Though remote communications has been mentioned in several contexts here, a brief treatment of its benefits is in order. For monitoring, reporting and alarm call-out, the communications feature offers tremendous utility. In each of these areas the ability to conduct system interrogation without a site visit is a valuable asset. The opportunity to modify and fine-tune the software aspect of the interface is also important.

In summary, the need for global building functions continues to grow. We will see more integrated systems of this type as property owners strive for state-of-the-art facility management. The fact the owners and users continue to expand their systems and find new values in automation confirms that these systems have a place in buildings of the future.

Automation has tremendous value in the information decade and, if past actions are an indicator, professionals in every segment of this industry will test the limits of their imaginations for new applications. One of the particularly rewarding aspects of this application is that it gave professionals from various disciplines the opportunity to be successful in a team approach. The bottom line, though, is that this application can contribute thousands of dollars a year to the profitability of each retail store. This is an automation benefit which is certain to attract the attention of any chief executive officer.

Appendix A

Software Dictionary

This dictionary was prepared by Honeywell to explain the terminology used when digital computers and software are employed in building control applications. It was written for those with experience in building control applications but with little or no computer experience. An attempt has been made to keep the definitions short and nontechnical.

Author of the Software Dictionary is Jack Vitelli, who is with the Honeywell Commercial Division.

A

ABORT—Premature termination of a program during execution.

ACCESS, DIRECT—Retrieval of information from memory where retrieval time is independent of the location in memory. Same as random accesss.

ACCESS, RANDOM—See Access, Direct.

ACCESS, SERIAL—Retrieval of information from memory where retrieval time is dependent on its location in memory. Information on magnetic tape requires serial access.

ACCESS TIME—The time required for the CPU to retrieve information from a memory device. A measure of the speed of a memory unit. Also called memory access time. Typical access times are as follows:

TTL RAM	60 ns
MOS RAM	300 ns
Core	500 ns
Bubble	3 ms
Fixed Head Disk	8 ms
Moving Head Disk	50 ms
Floppy Disk	100 ms
Cassettes	10 s
Tapes	10 s

ACCUMULATOR—A memory location (register) found in the arithmetic and logic unit of a digital computer. Used to accumulate and store intermediate arithmetic and logical results.

ACOUSTIC COUPLER—A device that allows digital information to be transmitted over voice-grade telephone lines. A type of modem.

ADA—A high-level language whose development is directed and funded by the U.S. Department of Defense. A language for writing real-time, command and control software. Likely to be a very popular language in the late 1980s and 1990s.

ADDRESS—A label or number specifying where a unit of information is stored in the computer's memory.

ADDRESS, ABSOLUTE—A reference to a memory location that has a fixed displacement from memory location zero.

ADDRESS, INDIRECT— An address that specifies a memory location that contains either an absolute address or another indirect address.

ADDRESS, RELATIVE—An address to which a number must be added in order to find the absolute address.

ADDRESS, SYMBOLIC—A label, alphabetic or alphanumeric, used to specify a storage location in the context of a particular program.

ALGOL—ALGOrithmic Language. A high-level language which is used primarily in scientific applications. Has gained much wider acceptance in Europe than in North America.

ALGORITHM—A prescribed set of well-defined rules or processes for the solution of a problem in a finite number of steps.

ALGORITHM, CONTROL—A calculation method that produces a control output by operating on an error signal or a time series of error signals.

ALGORITHMIC—A method of problem solving utilizing well-defined procedures or algorithms.

ALPHA—Alphabetic characters.

ALPHANUMERIC—Alphabetic, numeric and usually other characters such as punctuation marks.

ALU—See Arithmetic and Logic Unit.

ANALYST—A person who defines problems and develops algorithms and procedures for their solution.

AND—A Boolean operator which is true only when all inputs are true. Equivalent to a circuit of contacts arranged in series.

ANSI—American National Standards Institute. An organization which defines standards for computer languages, protocols, etc.

ARGUMENT—A value which is input to a subroutine or function.

ARITHMETIC AND LOGIC UNIT—A set of electronic circuits in the CPU which performs all arithmetic and logical operations. The ALU.

ARITHMETIC, FIXED POINT—Computer arithmetic in which the operands and results of all arithmetic operations must be properly scaled because the computer does not consider the position of the radix point. Contrasted with floating point arithmetic.

ARITHMETIC, FLOATING POINT—Computer arithmetic which automatically accounts for the location of the radix point in the operands and results.

ARRAY—A table of data arranged in a meaningful pattern.

ARTIFICIAL INTELLIGENCE—The capability of a machine to perform human-like intelligence functions such as learning, adapting, reasoning and self-correction. See Control, Adaptive.

ASCII— American National Standard Code for Information Interchange. The standard code, using a coded character set consisting of 7-bit coded characters (8 bits including parity check), used for information interchange among data processing systems, communications systems, and associated equipment. The ASCII set consists of the upper and lower case English alphabet, numerals, special symbols and 32 control codes.

ASSEMBLER—A program that translates assembly language into machine language.

ASSEMBLY LANGUAGE—See Language, Assembly.

ASYNCHRONOUS—Without regular time relationships. As applied to program execution, unpredictable with respect to time or instruction sequence. Opposite of synchronous.

ASYNCHRONOUS COMMUNICATION—Data transmission where the time interval between characters is allowed to vary.

ASYNCHRONOUS COMPUTER—One in which operations are not all timed by a master clock. The signal to start an operation is provided by the completion of the previous operation.

AUTOMATION—Automatically controlled operation of an apparatus, process or system by mechanical or electronic devices that take the place of human observation, effort and decision.

B

BABBAGE, CHARLES—The British Mathematician (1792-1871) whose "analytical engine" anticipated the digital computer by more than a centrury.

BACKGROUND—A partition of main memory used for processing batch jobs. The background is normally assigned the lowest priority in a multiprogramming environment. See also Foreground and Processing.

BACKGROUND PROCESSING—See Processing, Background.

BASIC—Beginners' All-Purpose Symbolic Instruction Code. A relatively easy-to-use high-level computer language that comes with many small and personal computer systems. Usually an interpretive language.

BATCH PROCESSING—See Processing, Batch.

BAUD—A Baud is a signal change in a communication link. One signal change can represent one or more bits of information depending on type of transmission scheme. Simple peripheral communication is normally one bit per Buad. E.g., Baud Rate = 1200 Baud/sec is 1200 bits/sec if one signal change = 1 bit.

BCD—Binary Coded Decimal. A method of representing each figure in a decimal number by a four bit binary number.

BENCHMARK—A standard program used to evaluate the performance of computers.

BINARY—The basis for calculations in all digital computers. This two-digit numbering system consists of the digits 0 and 1, in contrast to the ten-digit decimal system.

BINARY CODE—See Code, Binary.

BINARY CODED DECIMAL—See BCD.

BINARY DIGIT—See Bit.

BINARY NUMBER—A number written in the base 2 number system. When reading the number from right to left, each digit is understood to be multiplied by a progressively higher power of 2. The digits are often referred to as the 1's, 2's, 4's, 8's and 16's positions (etc.) 0100101 (binary) equals 37 (decimal).

BIPOLAR—A fundamental type of integrated circuit based on conventional PNP or NPN transistors formed from layers of silicon with different electrical characteristics.

BIT—Binary Digit. A single character in a binary number. A bit has two possible values, a 1 or a 0.

BIT, PARITY—A binary digit appended to an array of bits to make the sum of all the bits always odd or always even.

BIT-SLICE—A microprocessor architecture in which the basic elements of the microprocesscr are divided among several identical chips that can be linked in parallel, the total number of chips depending on the length of word the user wants to process.

BLOCK—A group of contiguous words, characters or records handled as a unit when transferring information between main and secondary memory.

BLOCK STRUCTURE—The characteristic of a program which has been segmented into logically closed groups (blocks) of language statements. Programming languages which force this type of structure are called block structured languages. Examples are Pascal, ALGOL, PL/1, Delta FORTRAN and Ada.

BOOLEAN ALGEBRA—A process of reasoning, or a deductive system of theorems using a symbolic logic, and dealing with classes, propositions, or on-off circuit elements. Named after George Boole, famous English mathematician (1815-1864).

BOOTSTRAP—A technique for loading the first few instructions of a routine into memory then using these instructions to bring in the rest of the routine.

BOOTSTRAP LOADER—See Loader, Bootstrap.

BPS—Bits Per Second.

BRANCH INSTRUCTION—A computer instruction that directs the computer to choose between alternate subprograms depending upon whether a specified condition is true or false at the time of execution of the program. Same as branch statement, conditional branch and conditional jump.

BREAK—A user action, initiated by pressing the break or interrupt key, that interrupts a running program so that commands can be entered, the program can be halted temporarily, or termination of the program can be effected.

BREAKPOINT—A point in a computer program at which conditional interruption (to permit visual check, printing out, or other analyzing) may occur. Breakpoints are usually used in debugging operations.

BUFFER—Memory area in a computer or peripheral used for temporary storage of information that has just been received. The information is held in the buffer until the computer or device is ready to process it.

BUG—A programming error or a hardware malfunction.

BUNDLED—A software pricing technique where software is provided in the purchase price of a hardware device on which the software executes.

BURN—To enter instructions and data permanently into PROM or EPROM.

BUS—A set of parallel conducting paths connecting various hardware units of a computer.

BUS, BIDIRECTIONAL—A data bus in which digital information can be transmitted in either direction.

BYTE—A group of 8 binary digits (bits) usually operated on as a unit by a digital computer. Memory size is sometimes given in the number of bytes and sometimes in the number of words.

C

C—A somewhat structured high-level language developed as the systems programming language and designed to optimize program run time, size and efficiency.

CACHE—A buffer type of high speed memory filled at medium speed from main memory with instructions and data most likely to be needed next by the CPU.

CALL—A program instruction that initiates execution of a subroutine.

CCDs—Charge Coupled Devices. Micro electronic circuit elements that can serially move quantities of electrical charge across the surface of a semiconductor. When used as computer memory, CCDs offer a very small, volatile, reliable memory with very low power dissipation per bit. Used primarily in signal processing.

CENTRAL PROCESSING UNIT—The hardware part (CPU) of a computer which directs the sequence of operations, interprets the coded instructions, performs arithmetic and logical operations, and initiates the proper commands to the computer circuits for execution. The arithmetic and logic unit (ALU) and the control unit of a digital computer. Controls the computer operation as directed by the program it is executing.

CHAINING—The process of having one program initiate or transfer control to another program.

CHANNEL—A path along which communications may be sent.

CHARACTER STRING—See String.

CHECKPOINT—A point during execution, at which information about a program is recorded which will permit subsequent resumption of processing from that point.

CHECK SUM—A method of detecting errors when a block of information is being loaded into a computer from a magnetic medium. The checksum is the sum of the numerical values of the bytes in the block. As the block is loaded, the checksum is computed. After the loading is complete, this value is compared with the checksum value that was placed on the medium when it was generated. If the two are equal, the information is assumed to have been loaded without error.

CHILLER OPTIMIZATION—An energy management application program that resets chilled and condenser water temperatures, and selects and loads chillers to provide necessary cooling at minimum cost.

CHILLER PROFILE—A program which provides historical output of measured and calculated values associated with a chiller plant but does not provide chiller control or chiller optimizing.

CHIP—A small piece of silicon impregnated with impurities in a pattern to form transistors, diodes, and resistors. Electrical paths are formed on it by depositing thin layers of aluminum or gold.

CIL—Control Interpreter Language. An interpretive language which enables the user to independently modify existing software applications or write entirely new control sequences. A variety of CIL operators allow control logic and arithmetic expressions to be programmed in easy-to-learn terms.

CLEAR—To place a memory device into a prescribed state, usually that denoting zero or blank.

CLOSED LOOP—A method of control in which feedback is used to link a controlled process back to the original command signal.

CMOS—Complementary MOS. A combination of PMOS and NMOS transistors on a single chip of silicon which results in an integrated circuit as fast as NMOS but consuming less power.

COBOL—Common Business Oriented Language. A high-level language which is used primarily in business applications.

CODE—The symbology used to represent computer instructions and data. To write a program.

CODE, BINARY—A code or language where information is represented by groups of 1's and 0's. The native language of every digital computer. Synonymous with machine code.

CODE, HEXADECIMAL—A computer code whose human understandability is between machine language and assembly language. Utilizes the integers 0 through 9 and the letters A through F, to represent groups of four binary digits.

CODE, INTERMEDIATE—A code developed by a computer which is not machine code and therefore not directly executable by a specific computer. At the time of execution of the program, intermediate code is typically executed by an interpretor on the target machine. Many recent high-level language compilers for microprocessors produce an intermediate code.

CODE, MICRO—A set of control functions performed by the instruction decoding and execution logic of a computer system.

CODE, OBJECT—The code produced by a compiler or assembler, and requiring further processing by a linker to produce binary or machine code.

CODE, RELOCATABLE—Machine code that can be automatically modified by a program called a relocating loader to occupy any position in main memory.

CODE, SOURCE—The code or language used by the programmer when the program was written. Code that must be processed by a compiler, assembler, or interpreter before it can be executed by the computer.

COMFORT-FAIRNESS SHED/ADD—A power demand shed/add algorithm which sheds and adds loads serving areas of a building based on the deviation of space temperature (or humidity) from a specified comfort level. Smallest deviation shed first.

COMMAND LANGUAGE—A set of commands used by an operator to control the overall operation of a computer system.

COMMENT—A statement of explanation within a program which has no effect on the operation of the computer at the time of program execution.

COMMUNICATION PROCESSOR—See Processor, Front-End.

COMPILE—To translate a high-level language program into a machine language or intermediate language program.

COMPILER—A program that translates a specific high-level language into a machine language or intermediate language. Each high-level language statement is usually translated into several machine instructions.

COMPILER LANGUAGE—See Language, High-Level.

COMPREHENSIVE SCHEDULER—A software routine that recognizes the occurrence of time, hardware, software or human events and causes preprogrammed actions to take place based on these events.

COMPUTER—A machine designed to receive data, perform operations on this data according to a program stored in its memory, and then produce a resultant output.

COMPUTER ANALOG—A computer which represents variables by physical analogies. A computer that solves problems by translating physical conditions such as flow, temperature, pressure or voltage into an electrical quantity and uses electrical equivalent circuits as an analog for the physical phenomenon being investigated.

COMPUTER. DIGITAL—A computer which processes information represented by combinations of discrete or discontinuous data, as compared with an analog computer for continuous data.

COMPUTER, HOST—The computer on which the program executes.

COMPUTER, MICRO—See Microcomputer.

COMPUTER, MINI—See Minicomputer.

COMPUTER SCIENCE—The entire spectrum of theoretical and applied disciplines for the development and application of computers.

COMPUTER, TARGET—The computer on which the software under development will execute operationally.

CONCATENATE—To unite in a series. To link together. To chain.

CONCURRENT PROCESSING—See Processing, Concurrent.

CONDITIONAL BRANCH—See Branch Instruction.

CONDITIONAL JUMP—See Branch Instruction.

CONTENTION—Rivalry for use of a system resource, such as main memory.

CONTROL, ADAPTIVE—A control algorithm or technique where the controller changes its control parameters and performance characteristics in response to its environment and experience. See self-adapting and self-learning.

CONTROL, CASCADE—An algorithm or method of control in which various control units are linked in sequence, each control unit regulating the operation of the next control unit.

CONTROL, DIRECT DIGITAL—A control loop in which a digital controller periodically updates the process as a function of a set of measured control variables and a given set of control algorithms.

CONTROL, DISTRIBUTED—Distribution of operational control, processing and data to controllers (usually digital computers) throughout a network of controllers connected by a common transmission system. Each controller can maintain independent local loop control if communication is lost with other controllers in the network. Distributed Control implies distributed processing and distributed data. Delta systems employ distributed control.

CONTROL, INTEGRAL—A control algorithm or method in which the final control element is moved in a corrective direction at a rate proportional to the deviation (error) of the controlled variable until the controller is satisfied or until a movement in the other direction is called for. Also called proportional speed floating control.

CONTROLLER—A device capable of measuring and regulating by receiving a signal from a sensing device, comparing this data with a desired value and issuing signals for corrective action. A digital computer is a type of controller.

CONTROLLER, PROGRAMMABLE—A solid state controller with a limited programming capability.

CONTROL, PI—Proportional/Integral Control. A control algorithm that combines the proportional (proportional response) and integral (reset response) control algorithms. Reset response tends to correct the offset resulting when proportional control alone is used. Also called proportional-plus-reset control and two-mode control.

CONTROL, PID—Proportional/Integral/Derivative Control. A control algorithm that enhances the PI Control algorithm by adding a component that is proportional to the rate of change (derivative) of the deviation (error) of the controlled variable. This term serves to anticipate future error under current conditions. Also called three-mode control.

CONTROL, PROPORTIONAL—A control algorithm or method where the final control element moves to a position proportional to the deviation of the value of the controlled variable from the set point.

CONTROL UNIT—Part of the CPU. It directs the sequence of operations, interprets coded instructions, and initiates proper commands to computer circuits to execute instructions. In the human body, the brain is the control unit.

CONVERSATIONAL—A mode of computer operation where a human communicates interactively (directly) with the central processing unit by issuing commands, responding to requests for input and interpreting results. The computer presents choices to the operator from which is chosen a desired function. See Menu Penetration.

COPY—To reproduce a program or data while leaving the original information unchanged.

CORE—Sometimes used synonymously with main memory since main memory was at one time exclusively the core type. See Memory, Core.

CORE RESIDENT—A program permanently located in main memory.

COUNTER—A register or memory location used to count the number of times a certain event occurs.

CPS—Characters per second or cycles per second.

CPU—See Central Processing Unit.

CRASH—An ungraceful system shutdown caused by a hardware or software malfunction.

CROSS-ASSEMBLER—An assembler that translates assembly language programs for computer A into machine language programs for computer A although the assembler runs on computer B.

CROSS-COMPILER—A compiler that runs on a machine other than the one for which it is designed to compile code. A compiler which runs on computer A and produces machine or intermediate code that will execute on computer B.

CROSS SOFTWARE—Software that lets users develop programs for a target computer on a host computer.

CRT—Cathode Ray Tube terminal. A computer terminal that displays its output on a television-like screen. Also video terminal. May be color or black and white.

CURSOR—A symbol on the display of a video terminal that indicates where the next character is to be located.

CURVE PLOTTING—A program which plots system data in the form of graphic curves and bar charts.

CYCLE STEALING—A technique that allows an external device to temporarily disable processor control of the bus. This, in turn, allows the device to access main memory.

CYCLE TIME—The time interval between the call for, and the delivery of, information from a memory unit. A method for implementing direct memory access (DMA).

D

DATA—The raw information accessible to, operated on and produced by the computer.

DATABASE—A large collection of records stored on a computer system from which specialized data may be extracted, organized, and manipulated by a program. Any organized and structured collection of data in memory. See Database Management System.

DATABASE LANGUAGE—See Language, Database.

DATABASE MANAGEMENT SYSTEM—A software program that provides the functions of storing, updating and retrieving information stored in a database where multiple users/computers can instantaneously access the information. See Total.

DATA DESCRIPTION LANGUAGE—See Language, Data Description.

DATA FILE—A file of raw data in a format understood and used by a program.

DATA MANIPULATION LANGUAGE—See Language, Data Manipulation.

DBMS—See Database Management System.

DDC—See Control, Direct Digital.

DEAD BAND—A temperature range over which no heating or cooling energy is supplied. Opposite of overlap.

DEBUG—To detect, locate, and remove mistakes from a routine or malfunction from a computer. Synonymous with troubleshoot.

DEBUG PROGRAM—A utility program that provides an on-line capability to inspect, control and debug another program during execution.

DECIMAL NUMBER—A number written in the base 10 number system. When reading the number from right to left, each digit is understood to be multiplied by a progressively higher power of 10. The digits are often referred to as the 1's, the 10's, and 100's positions (etc.).

DECISION TABLE—A table of all contingencies that are to be considered in the description of a problem, together with the actions to be taken. Decision tables are sometimes used in place of flowcharts for problem description and documentation.

DEFAULT VALUE—The value assumed by a variable in a program unless another value is specified by the user.

DEGRADATION—System operation continues at a reduced level of service.

DEGRADATION, GRACEFUL—See Failsoft.

DELIMITER—A character that denotes the beginning and end of a character string and, therefore, cannot be a member of the string.

DEMAND CONTROL— See Power Demand Control.

DEVICE—A computer peripheral or an electronic component.

DGP—Data Gathering Panel. An electronic panel used for data concentrating, preprocessing (A/D conversion) and parallel to serial conversion (and vice versa) allowing two-way serial transmission between remote locations and a central CPU.

DGP, INTELLIGENT—A microprocessor-based DGP providing the basic DGP functions as well as additional functions such as lockout control, run-time accumulation, local point control, and local execution of application software such as duty cycling, load reset, optimum start/stop and event initiated programs.

DIAGNOSTIC MESSAGE—An output message signaling a problem and providing a hint as to its origin.

DIAGNOSTIC ROUTINE—See Routine, Diagnostic.

DIAGNOSTICS, ON-LINE—Diagnostic messages that are output to an operator or user console. Executed while normal system functions continue.

DIAGNOSTIC TEST—Executing a diagnostic routine.

DIP—Dual In-line Package. A type of integrated circuit packaging. It is characterized by a rectangular shape and by pins that point downward. The pins are arranged on two sides of the rectangle.

DIRECT DIGITAL CONTROL—A control loop in which a digital controller periodically updates the process as a function of a set of measured control variables and a given set of control algorithms.

DISK—A secondary memory device in which information is stored on one or both sides of a magnetically sensitive, rotating disk. The disk is rotated by a disk drive and information is retrieved/stored by means of one or more read/write heads mounted on movable or fixed arms. Disks are rigid (hard) or flexible (floppy). Disk memory has a much faster access time than magnetic tape but is slower than semiconductor memory.

DISK, CARTRIDGE—A disk memory unit where one or more of the disks are easily removed and replaced by another.

DISK DRIVE—The part of a disk memory unit on which disks are mounted and which rotates the disks.

DISK, DUAL FLOPPY—Two floppy disk drives packaged together.

DISKETTE—A floppy disk. See Disk, Floppy.

DISK, FIXED HEAD—A disk memory unit on which a separate read/write head is provided for each track of each disk surface. A typical disk has about 800 tracks per side. Access time on a fixed head disk is typically about 10 milliseconds and much faster than access time for a moving head disk which is typically 25 to 50 milliseconds.

DISK, FLOPPY—A circular, flexible material with a magnetic film on both sides. Digital information is stored on one or both sides of the floppy disks in concentric circles or tracks. The term "floppy" is used because the disk is soft and bends easily. The disk is usually contained in a rigid, protective envelope. There are two sizes of floppy disks. They are 8 inches and 5¼ inches in diameter.

DISK, HARD—Disk memory that uses rigid disks rather than flexible disks as the storage medium. Hard-disk devices can generally store more information and access it faster. Cost considerations, however, currently restrict their usage to medium and large-scale applications.

DISK MEMORY UNIT—A secondary memory device often used in real-time control systems. A disk memory unit consists of one or more disks stacked on top of one another. The disks are rotated continuously at a uniform speed. Information is stored on the magnetic disks in concentric tracks. A typical disk surface can record between 1M and 100M bytes of information.

DISK, MOVING HEAD—A magnetic disk unit that has one head per surface. The read/write head moves horizontally to locate information on about 800 concentric tracks on each disk. The movable head can be completely retracted so that a disk pack can be removed.

DISK PACK—A set of magnetic disks mounted on a shaft and designed so that the pack can be removed from the disk drive unit and replaced with another disk pack.

DISK, WINCHESTER—A hard, fixed head disk memory unit which is hermetically sealed to improve reliability and increase its useful life. A separate device is required for loading and dumping software from the disk because it is not removable.

DISTRIBUTED CONTROL SYSTEM—A distributed system of the highest order. Operational control, processing and data are distributed to computers throughout the system. Little functionability is lost in any processor on the network if communication is lost with other processors. Distributed control usually implies distributed processing and distributed data. Delta is a distributed control system.

DISTRIBUTED PROCESSING—A sharing of the processing (CPU) load by all computers in a distributed system. See Control, Distributed.

DISTRIBUTED SYSTEM—A collection of autonomous computer systems all connected by a communication mechanism where data, processing and/or control is distributed throughout the system. Control is the more difficult to distribute followed by processing and then data. See Control, Distributed.

DMA—Direct Memory Access. A process where blocks of data can be transferred between main memory and secondary memory (such as a disk or diskette) without processor intervention.

DOS—Disk Operating System. An operating system which utilizes direct access secondary memory.

DOUBLE PRECISION—An optional computer arithmetic where twice as many digits are handled in arithmetic operations than are normally handled by the computer. Increased accuracy results.

DOWNLINE LOAD—The electronic transfer of programs and data files from a central computer, with secondary memory devices, to remote, distributed computers in a distributed system. Transfer is made over the distributed system's communication network.

DOWNTIME—The period during which a computer is not operating or is malfunctioning because of a machine fault or failure. Distributed control reduces downtime.

DRIVER—A program which controls the use of (drives) a peripheral device.

DRUM, MAGNETIC—A secondary memory device in which information is stored on the surface of a magnetically sensitive, rotating, right circular cylinder.

DUMP—To copy the contents of all or part of one memory unit to another. Usually to copy main memory onto a secondary memory device such as a tape or disk.

DUTY CYCLING—An energy management application program which reduces the consumption of electrical energy by cycling equipment so that minimum operating time is used to maintain comfortable, environmental conditions. Advanced duty cycling programs allow temperature compensation feedback to insure comfort.

DYNAMIC ALLOCATION—Providing immediate resources (memory space, I/O devices, etc.) to a program in response to a demand by the program.

DYNAMIC MEMORY—See Memory, Dynamic.

DYNAMIC PROGRAM LOADING—The process of loading a program into main memory upon the demand for that program by an executing program.

E

EAROM—Electrically Alterable ROM. A type of memory that combines the characteristics of RAM and ROM. It is nonvolatile (like ROM) but can be written into by the processor (like RAM). The EAROM, however, has a substantially longer writing time (currently about 2 microseconds vs 400 nanoseconds) as well as a limited number of writes (about 1,000,000) before the chip can no longer be reprogrammed.

EBCDIC—Extended Binary Coded Decimal Interchange Code. A specific code using eight bits to represent a character. Contrast with ASCII.

EDAC—See Memory, EDAC.

EDIT—To modify programs, data files, text files and database information by adding, deleting and changing characters, lines, records, words, etc.

EDITOR—A utility program which performs editing operations on programs and data.

EDP—Electronic Data Processing. The transformation of raw data into useful information by electronic equipment.

EIP—See Event Initiated Program.

EMPIRICAL—Based on experience rather than deductive logic.

EMULATION—The use of programming techniques and special machine features to permit a computing system to behave exactly like another system.

EMULATOR—A program that allows one processor to simulate the instruction set of another processor.

EMULATOR, IN-CIRCUIT—Replacing the microprocessor on a circuit board with a plug connected to a test tool (usually a microcomputer) which simulates the real-time operation of the microprocessor in a non-real-time environment.

ENCODE—To put into code.

ENERGY MANAGEMENT SOFTWARE—Refers to those programs which reduce energy consumption in a building. They range from very simple to extremely complex.

ENTHALPY—A measure of the total (latent and sensible) heat content of the air.

ENTHALPY CONTROL—An energy management application program which reduces cooling cost by selecting either the outdoor air or return air stream to be cooled which has least enthalpy.

EPROM—Electrically Programmable ROM. A nonvolatile semiconductor memory that can be erased and reprogrammed. Programming is done off-line with an EPROM or PROM burner. Erasure is accomplished by exposure to ultraviolet light.

ERROR ROUTINE—A program which automatically initiates corrective action when a hardware or software error is detected.

EVENT INITIATED PROGRAM—A program which issues specified commands to any number of system points based on a system event such as a hardware change-of-status, absolute or elapsed time, an operator command or a command from another program. Called EIP.

EXECUTE—To perform a computer instruction or run a program.

EXECUTION TIME—The time required for the CPU to decode and execute a computer instruction.

EXECUTIVE—A routine designed to organize and regulate the flow of work in a computer system by initiating and controlling the execution of other programs. A principle component of most operating systems. Synonymous with Supervisory Routine, Executive Routine and Supervisor.

EXECUTIVE, REAL-TIME—Multitasking executive system that handles all aspects of priority scheduling, timing, interrupt servicing, input-output control, intertask communications, and all necessary queuing functions.

F

FAILSOFT—In the event of a hardware failure, to fall back to a degraded mode of operation rather than let the system fail completely with no response to the user. Implemented by a system program. Also called graceful degradation.

FALLBACK—A mode of operation where manual or special program procedures are used to maintain some level of performance.

FAULT-TOLERANT—Pertaining to software or systems which will still execute properly even though parts may fail.

FEEDBACK—The signal (or signals) fed back from a controlled process to denote its response to the command signal.

FETCH—To retrieve data from memory.

FIBER OPTICS—A communication technique where information is transmitted in the form of light over a transparent material (fiber) such as a strand of glass. Advantages are noise free communication not susceptible to electromagnetic interference.

FIELD—A specified area of a record in which a particular item of information is stored. For example, the second field of a system point record is its current alarm status.

FILE—A collection of one or more records treated as a unit. The records may consist of data or program instructions.

FILE, DATA—A file of raw data in a format understood and used by a program.

FILE, PARAMETER/RESULT—A file containing data which can be accessed and changed by a program without the need for file input/output statements.

FILE, SEQUENTIAL—A file in which elements are stored serially.

FIXED INTERVAL IDEAL RATE—A power demand control algorithm which attempts to return power consumption to an ideal rate slope by the end of each sample period within the utility company's demand interval.

FIXED-INTERVAL PREDICTIVE—A power demand control algorithm which attempts to return power consumption to a specified target by the end of the utility company's demand interval.

FLAG—A bit whose state signifies whether a certain condition has occurred.

FLAT PACK—A type of integrated circuit packaging in which the pins extend outward, rather than pointing down as on a DIP.

FLIP-FLOP—A circuit that changes its logical state when signaled to do so by another device.

FIRMWARE—Microprogrammed logic circuits resident in a computer, peripheral controller, terminal or other memory unit. Software that is resident in ROM or PROM is sometimes referred to as firmware.

FLOWCHART—A diagram representing the logic of a computer program. An obsolete method of documenting software.

FLOWCHART, MACRO—A flowchart that provides an overview of the program logic but not all the detail.

FOREGROUND—The environment or main memory partition in which high priority, real-time control programs are executed. Contrast with background.

FOREGROUND PROCESSING—See Processing, Foreground.

FORTH LANGUAGE—A high-level language, usually implemented on microcomputer systems, whose major feature is that the language is easily extended by the user through user-defined operators (procedures, functions or commands).

FORTRAN—FORmula TRANslator. The first high level language. Introduced in the early 1950s. Facilitates the programming of mathematical and algebraic operations. FORTRAN IV is an advanced version introduced in 1964. Widely used in scientific and engineering applications.

FRONT-END PROCESSOR—See Processor, Front-End.

FULL-DUPLEX—A communication facility providing simultaneous transmission and reception.

FUNCTION—A program or routine which returns a single value when its execution is requested in a program. For example, square root, absolute value, Fahrenheit to centigrade conversion, etc. Compare with subroutine.

FUNCTION, INTRINSIC—A function built into a high-level language as a standard feature. Examples are square root, absolute value, enthalpy and the trigonometric functions.

G

GATE—A circuit having one output and several inputs, the output remaining unenergized until certain input conditions have been met. A circuit that performs a Boolean logic operation.

GENERATOR—A program that creates another program or data.

GENERATION, SYSTEM—A process which creates a particular and uniquely specified operating system. System generation combines user-specified options and parameters with manufacturer-supplied, general-purpose programs to produce an operating system configured specifically for the user's application.

GLOBAL—Pertaining to data or parameters accessible throughout an entire program or by all controllers in a distributed control system.

GRAPHIC, BLOCK—Referring to a CRT unit where graphics are constructed from solid rectangles which occupy one character location.

GRAPHIC, FULL—Referring to a CRT display unit (usually color) where each individual pixel of a graphic is controllable with software. Individual pixel control permits more information, more detailed graphics and smooth curves to be presented on the CRT screen. Contrast with semi-graphic and block graphic.

GRAPHIC, SEMI—Referring to a CRT display unit where graphics are constructed from a fixed character set established in PROM. The character set is made up of rectangular, two color arrays of pixels (5 X 7, 7 X 9, etc.). A standard CRT terminal is a type of semi-graphic unit where the character set is the upper/lower case alphabet, the numbers and a few other characters (%, *, $, etc.). Also called character graphic.

GRAPHICS—The use of diagrams or other graphic means to present operating data, curve plots, answers to inquiries and other computer output.

GRAPHICS TERMINAL—A CRT terminal capable of displaying user-programmed graphics.

GROUP—A collection of points. See Point, Physical and Point, Pseudo.

GROUP, LOGICAL—A collection of points whose number and order are established by software and hence not constrained by the hardware configuration.

GROUP, PHYSICAL—A collection of points controlled by a single DGP whose number and order are determined by the hardware configuration of the DGP.

H

HALF-DUPLEX—A communication facility providing both transmission and reception, but not simultaneously.

HANDLER—A routine which controls communication with an external device.

HANDSHAKING—A preliminary procedure performed by modems and/or terminals and computers to verify that communication has been established and can proceed.

HANG-UP—An unwanted halt in the execution of a program.

HARD COPY—Output printed on a permanent medium, such as paper.

HARDWARE—The physical components of the computer processing system. For example, mechanical, magnetic, electrical or electronic devices. Contrast with software.

HEADER—The first part of a file or message containing identification data and characteristics of the file or message.

HEURISTIC— Exploratory methods of problem solving in which solutions are discovered by evaluation of the progress made toward the final result. Problem solving by intuitive trial-and-error. Contrast with algorithmic.

HEXADECIMAL CODE—See Code, Hexadecimal.

HEXADECIMAL NUMBER—A number written in the base 16 number system. The first 10 numbers in the system are the integers 0 through 9 while the last 6 "numbers" are the letters A, B, C, D, E, F. Also called hex number.

HIERARCHY—A specified rank or order. Software hierarchy is often based on which programs are prerequisites of others.

HIGH-LEVEL LANGUAGE—See Language, High-Level.

HIPO DIAGRAM—Hierarchical Input-Process-Output. Each program module is defined by specifying its required input data, what it does (process) and the resultant output.

HOST COMPUTER—The computer on which the program executes.

HYSTERESIS—The lag in an instrument's or process response when a force acting on it is abruptly changed.

I

IC—See Integrated Circuit.

IF-THEN-ELSE—A program statement often used in modern high-level language to provide a closed branch in the program. When the IF statement is true, the THEN procedure is executed. Otherwise the ELSE procedure is executed.

IMAGE PROCESSING—See Processing, Image.

INDEXING—The process of establishing memory addresses by adding the value in an address field of an instruction to a value stored in a specified index register.

INITIALIZATION, SYSTEM—The process of loading the operating system into the computer and defining the processing environment.

INITIALIZE—To originate or establish the basic conditions or startup state.

INPUT—Information or data supplied to a computer for processing.

INPUT/OUTPUT—See I/O.

INQUIRY—Interrogation of a computer system initiated at a keyboard.

INSTRUCTION—An explicit command that tells the computer what to do. Control instructions govern the transfer of information within the machine as well as between the machine and the I/O devices. Arithmetic and logic instructions specify the arithmetic and logic operations to be performed on data.

INSTRUCTION SET—The repertoire of operations that a CPU can execute.

INTELLIGENT DEVICE—A device that contains a microprocessor.

INTELLIGENT DGP—See DGP, Intelligent.

INTERFACE—The bringing together, in an organized manner, of entities such as hardware, software, and a human.

INTERLEAVE—To insert segments of one program into another program to produce an effect of simultaneous execution by a single processor.

INTEGRATED CIRCUIT—A complex electronic circuit fabricated on a single chip of semiconductor material such as silicon.

INTERMEDIATE CODE—See Code, Intermediate.

INTERPRETER—A program (language processor) that translates and executes each high-level language statement or p-code statement before proceeding to the next statement. Contrast with a compiler which translates the entire high-level language program into machine language before any statement is executed. The BASIC language is most often translated by an interpreter.

INTERRUPT—The initiation, by hardware, of a routine intended to respond to an external (device-originated) or internal (software-originated) event that is either unrelated, or asynchronous with, the executing program.

INTERRUPT, VECTORED—An interrupt scheme where each interrupting device causes the operating system to branch to a different interrupt routine.

INTERRUPT STACKING—Recording the occurrence of one or more interrupts and allowing them to remain pending while processing continues.

I/O—Input/Output. A general term for the equipment used to communicate with a computer and the data involved in the communication.

I/O, PARALLEL—The simultaneous transfer of all bits in a byte or word over parallel conductors. Contrast with serial I/O.

I/O, SERIAL—Data transmission in which the bits are sent one by one over a single wire.

ITERATE—To repeat a series of computer instructions until some condition is satisfied. Also see Loop.

J

JUMP—A computer instruction that causes the processor to get its next instruction from a specified address in memory.

JUMP, ABSOLUTE—A jump instruction that causes the processor to get its next instruction from an absolute address in memory.

JUMP, CONDITIONAL—See Branch Instruction.

JUMP, RELATIVE—A jump instruction that causes the processor to get its next instruction from a memory address determined by adding a given number to the address at which the last instruction resided.

K

K—A unit for measuring the capacity of a computer memory. 1K is equal to 2^{10} or 1,024 units. 4K is 4,096 units. Memory size is usually measured in words or bytes.

L

LABEL—One or more characters used to identify an item of data.

LANGUAGE, ALGORITHMIC—A language in which mathematical algorithms are easily written. ALGOL is an algorithmic computer language.

LANGUAGE, ASSEMBLY—A symbolic, nonnumeric language in which each mnemonic instruction corresponds to one line of the source program and to one machine instruction.

LANGUAGE, COMPILER—See High-Level Language.

LANGUAGE, CONTROL INTERPRETER—See CIL.

LANGUAGE, DATABASE—A high-level language designed specifically for writing programs which retrieve and store information in a database. Called DBL.

LANGUAGE, DATA DESCRIPTION—A language for describing the named data components of a database and the relationships between these components. Called DDL.

LANGUAGE, DATA MANIPULATION—A high-level language used to communicate between the application program and the database. Relies on a host language like FORTRAN or COBOL. Called DML.

LANGUAGE, HIGH-LEVEL—A computer language that allows the programmer to write programs using English verbs, symbols and commands rather than machine code. Some common high-level languages are FORTRAN, COBOL, Pascal and BASIC. Each high-level language instruction corresponds to several machine instructions. Thus programs are easier to write, read and understand.

LANGUAGE, INTERPRETIVE—Any high-level language whose language processor is an interpreter.

LANGUAGE, MACHINE—Machine instructions in binary bit patterns that the processor can execute directly without additional interpretation. Contrast with high-level language or assembly language.

LANGUAGE, MANAGEMENT REPORT—See MRL.

LANGUAGE PROCESSOR—A program which translates a computer language (source code) into machine instructions (object code). Most language processors are categorized as compilers, interpreters or assemblers. They permit computer programs to be written in a language better understood by humans.

LANGUAGE, PROGRAMMING—A set of representations, rules and conventions used to convey instructions and data to a computer. Used to write computer programs. A programming language must be translated into machine language (see Language Processor) before it can be executed by the computer. Examples are FORTRAN, COBOL, BASIC, CIL.

LANGUAGE, SYSTEM PROGRAMMING—A language specifically suited for writing operating systems, compilers and database management systems but generally not application software.

LIGHT PEN—An input device used in conjunction with a video display. When the user touches the display screen with the light pen, the electronics associated with the pen will then be transmitted to the computer.

LINK—To establish communication between two or more programs in memory.

LINK EDITOR—A utility program that links one or more object programs into a single machine language program ready for execution, called a Load Module. Synonymous with linker.

LISP—LISt Processing. An interpretive language developed for the manipulation of character strings. Used to develop higher level languages.

LISTING, PROGRAM—A printed list of computer instructions, usually prepared by an assembler or compiler.

LITERAL—A symbol that names itself and that is not the name of something else. Contrast with label.

LOADER—A utility program that loads machine language programs and data into main memory from an input device. Also see Loader, Linking.

LOADER, BOOTSTRAP— A utility program, usually permanently resident in main memory, which enables other programs to load themselves.

LOADER, LINKING—A utility program that both links and loads into memory one or more object programs.

LOAD LEVELING—An algorithm that distributes the on/off cycles of equipment in each cycle period of a duty cycling program.

LOAD MODULE—A program unit that is discrete which has been compiled or assembled and linked. It is in machine language, and is directly executable by the computer. See Code, Binary.

LOAD RESET—An energy management application program which reduces the amount of energy used for heating and cooling by resetting the setpoint of the discharge temperature of the primary conditioner based on the load of greatest demand.

LOAD SHEDDING—See Power Demand Control.

LOCATION—A position in memory to store one or more computer words. Identified by an address.

LOGIC—As regards microprocessors, a mathematical treatment of formal logic using a set of symbols to represent quantities and relationships that can be translated into switching circuits or gates. Such gates are logical functions such as AND, OR, NOT, and many others. Each such gate is a switching circuit that has two states, open or closed. They make possible the application of binary numbers for solving problems. The basic logic functions electronically performed from gate circuits are the foundation of the often complex computing capability.

LOGIC GATE—A switching circuit that has two states, open or closed. These circuits make possible the application of binary numbers for solving problems. The basic logic gates (AND, OR, NOT) electronically performed from gate circuits are the foundation of the arithmetic and logic unit and of digital computing. See AND, OR, and NOT.

LOOP—A sequence of computer instructions that is executed repeatedly until a specified condition is satisfied. In automatic control, the path followed by command signals, which direct the work to be done, and feedback signals, which flow back to the command point to indicate what is actually being done.

LSI—Large Scale Integration. The process of integrating 1000 to 100,000 (not well defined) circuits or equivalent components on a single chip of semiconductor material. Up to 2000 logic gates or up to 16,000 bits on a chip. Microprocessors employ LSI technology.

LSI CHIP—An integrated circuit with LSI circuit/component density.

M

M—A unit for measuring the capacity of a computer memory. 1M is 2^{20} or 1,048,576 units. Memory size is usually measured in words or bytes. Short for mega or million.

MACHINE—Computer or CPU.

MACHINE CODE—See Code, Binary.

MACRO—A group of instructions that the programmer can define and name once at the beginning of a program and then invoke the execution of these instructions many times during the program by simply referencing the macro's name. This increases coding efficiency and readability of the program.

MACROASSEMBLER—An assembler that allows the programmer to define macros in assembly language.

MACROINSTRUCTION—An instruction in a source language that invokes the execution of a macro.

MACROPROCESSOR—A language processor whose source language instructions are equivalent to a specified sequence of machine instructions.

MAGNETIC CORE—See Memory, Core.

MAGNETIC DISK—See Disk.

MAGNETIC DRUM—See Drum, Magnetic.

MAGNETIC TAPE—See Tape, Magnetic.

MAINFRAME—The central processing unit, main memory and I/O interfaces of a computer.

MAINTENANCE, SOFTWARE—See Software Support.

MATRIX—A group of numbers organized on a rectangular grid and treated as a unit. The numbers can be referenced by their position in the grid.

MEDIA—The physical material on which data is recorded. Examples are paper tape, cards, magnetic disk, magnetic tape and bubble memory.

MEGA—When measuring computer memory, 1,048,576 units. Otherwise one million. Abbreviated M.

MEMORY—A device or section of the computer in which computer instructions and data can be stored and retrieved. Synonymous with storage.

MEMORY, BUBBLE—Bubble memories are made from magnetic materials called garnets. By applying a magnetic field, very small (2 to 5 microns) magnetized regions are formed which look like bubbles when viewed in polarized light under a microscope. Bubble memory access time is slower than semiconductor memory but offers nonvolatility, low power dissipation and no moving parts.

MEMORY CAPACITY—The maximum number of storage positions in the main memory of a computer.

MEMORY, CORE—A memory device made up of a matrix of thousands of ferromagnetic cores or rings strung together on a grid of wires. Individual cores are magnetized based on current direction in the wires. Through the 1960s most main memory was of this type. Core memory is sometimes used synonymously with main memory even though main memory is most typically of the semiconductor type in modern computers.

MEMORY, DEDICATED—Main memory locations reserved by the system for special purposes, such as interrupts and real-time programs.

MEMORY DUMP—See Dump.

MEMORY, DYNAMIC—A type of semiconductor memory which, unlike static memory, must be refreshed or recharged periodically to prevent loss of data. Also, memory which is dynamically allocated to tasks during program execution.

MEMORY, EDAC—Error Detection and Correction memory. A type of main memory which when read, attempts to correct any internally caused data error, thus improving total system reliability.

MEMORY, MAIN—The memory that the central processing unit accesses directly. Programs and data in secondary memory must be brought into main memory before they can be executed. Main memory is almost always the semiconductor type because of its fast access time. Also called core and main storage.

MEMORY MAP—A listing of all variable names, array names, constants and statement identifiers used by the program, with their relative address assignments. The listing includes all subroutines called and last location when called. A memory map is usually produced by a compiler, assembler, or linker.

MEMORY, NONVOLATILE—Memory which does not lose its stored information and data when power is lost.

MEMORY, OFF-LINE—Memory not under control of the CPU.

MEMORY, ON-LINE—Memory under control of the CPU.

MEMORY, PROGRAMMABLE—Memory that can be both read from and written into by the processor. Synonymous with RAM.

MEMORY, PROTECTED—Programmable memory that cannot be written into, usually on a temporary basis.

MEMORY, RANDOM ACCESS—See RAM.

MEMORY, READ-ONLY—See ROM.

MEMORY, READ-WRITE—Computer memory which can be both read from and written into.

MEMORY, SCRATCHPAD—Memory that is used temporarily during program execution to store data created and used by the program only during execution.

MEMORY, SECONDARY—Memory which is not directly addressable by the central processing unit. Information in secondary memory must be moved into main memory before it can be operated on by the CPU. Secondary memory is usually less expensive and has a slower access time than main memory. There is usually more secondary memory than main memory. Typical secondary memory devices are magnetic disks, tapes and drums.

MEMORY, SEMICONDUCTOR—A memory device whose storage medium is a semiconductor circuit. Commonly used as main memory because of its fast access time. Contrast with magnetic or core memory.

MEMORY, STATIC—Memory that does not require a refresh circuit to retain its information.

MEMORY, VIRTUAL—A conceptual extension of main storage achieved via a software or hardware technique which permits storage address references beyond the physical limitations of main memory. Virtual addresses are equated to real addresses during actual program execution.

MEMORY, VOLATILE—Memory which loses its stored instructions and data when power is lost.

MENU PENETRATION—A man-machine interface technique in which operator control of the system is accomplished by the selection of desired actions from tables or lists of alternatives (menus).

MESSAGE—A quantity of transmitted information that is physically continuous and processed as a unit.

MICROCODE—Software that defines the instruction set of a microprogrammable processor.

MICROCOMPUTER—Includes one or more microprocessors which perform logical and arithmetic functions under program control (firmware and/or software). Usually consists of one or more LSI chips, some ROM and some RAM. Word lengths range from 4 to 32 bits with 8 and 16 bit word lengths most common. Some I/O interface logic and a limited complement of I/O devices are offered.

MICROPROCESSOR—An LSI chip containing the arithmetic and logic unit and the control unit of a computer. It is simply the CPU of a digital computer placed on a chip. Also called an MPU.

MICROPROGRAM—A sequence of elementary microinstructions each corresponding to a computer hardware structure. The execution of a microprogram is usually initiated by a computer software instruction.

MICROPROGRAMMABLE PROCESSOR—A processor whose actual instruction set is not accessed by the programmer. Instead, another, more versatile, instruction set is simulated by microcode.

MICROSECOND—One millionth of a second (10^{-6} seconds). Abbreviated μs.

MILLISECOND—One thousandth of a second (10^{-3} seconds). Abbreviated ms.

MINICOMPUTER—Larger in size than a microcomputer and having a typical word length of 16 or 32 bits. Addressable main memory is typically from 64K to over 1M. Compared with a microcomputer, the minicomputer is typically characterized by higher performance, a richer instruction set, higher price, a variety of high-level languages, several operating systems and networking software.

MLCP—See Multiline Communications Processor.

MMI—Man-Machine Interface. A program which provides the human interface functions for a computer system. The driver for the primary operator terminals of the system.

MNEMONIC—An abbreviation for a computer instruction. For example, "jump-to-subroutine" might be represented by the mnemonic JSR.

MODEL—A mathematical or computational representation of a process, device or concept.

MODEM—A MODulation/DEModulation device that permits computers and terminals to communicate over telephone lines. Also called dataset.

MODULAR PROGRAMMING—The breaking up of a large program into small, self-contained modules. A modular program is more likely to meet the needs of the user because careful preparation and planning are needed in the program design stage. As the modules are implemented, they can be exhaustively tested, separately and together, thus ensuring that the modules do indeed conform to the user's requirements. In use, a modular program is easily adaptable because it is necessary to revise only the particular modules involved in the processing that requires later modification.

MODULARITY—Hardware modularity enables a user to start modestly and easily add new hardware units, memory capacity and terminals as added functionality is required. Software modularity means the ability to easily add new software modules to the system without scrapping existing software (operating system, applications) and without major system reconfigurations.

MODULE—A program unit that is discrete and identifiable for compiling, assembling and loading purposes. Also the input to and output from a single execution of a language processor. A source module, object module or load module. A software program, routine or subroutine.

MONITOR—A program that exercises primary control of the routines that comprise the operating system. A program that controls I/O and related functions. Used synonymously with Supervisor.

MOS—Metal Oxide Semiconductor. Refers to the three layers used in forming the gate structure of a field-effect transistor (FET). Refers to a fundamental process for fabricating integrated circuits of very high densities used in memories, microprocessors, watches and calculators.

MPU—Microprocessor Unit. See Microprocessor.

MULTILINE COMMUNICATIONS PROCESSOR—A programmable interface between a central processor and one or more communications devices. Can be programmed to handle specific communications devices.

MULTIPLEX—To interleave or simultaneously transmit two or more messages on a single channel.

MULTIPROCESSING—The employment of multiple interconnected CPUs to execute two or more different programs simultaneously.

MULTIPROGRAMMING—A technique for executing numerous programs simultaneously by overlapping or interleaving their execution. Permits more than one program to time share machine components.

MULTITASKING—An environment in which several separate but related tasks operate under a single program identity. Differs from multiprogramming in that common routines, data space and disk files may be used. May or may not involve multiprocessing.

N

NANOSECOND—One billionth of a second (10^{-9} seconds). Abbreviated ns.

NEST—To embed a subroutine or block of data within a larger routine or block of data.

NIGHT CYCLE—An energy management application program which maintains specified levels of temperature in the controlled space during unoccupied periods.

NIGHT PURGE—An energy management application program which causes the cool night air to be used to reduce the temperature in a building thereby reducing energy requirements during morning occupancy cool down.

NMOS—N-channel MOS. A MOS circuit in which electrons are the charge carriers. Faster than PMOS.

NOT—A Boolean operator which changes the input to its opposite value. Analogous to the set/reset function.

NS—Nanosecond. One billionth of a second.

NYBBLE—A group of four bits, or one-half of a byte.

O

OBJECT CODE—See Code, Object.

OBJECT PROGRAM—A program that has been translated into object code by a language processor (compiler, assembler). See Code, Object.

OCTAL—Referring to the base 8 number system or a base 8 number.

OFF-LINE—Describing a peripheral equipment state or user activity which is not under the control of the CPU.

OFFSET—The difference between the value or condition desired and that actually attained.

ON-LINE—Describing a peripheral equipment state or user activity which is under the control of the CPU.

OP CODE—Operation Code. A number or mnemonic that specifies the machine instruction to be executed.

OPEN LOOP—A method of control in which there is no self-correcting action for the error of the desired operational condition.

OPERAND—The input to or the result of a machine operation.

OPERATING SYSTEM—A main program that schedules and controls the execution of all other programs used by the computer. An organized collection of routines and procedures for operating a computer. These routines and procedures normally perform most of the following functions:

1) scheduling, loading, initiating and supervising the execution of programs,
2) allocating storage, input/output units and other facilities of the computer system,
3) initiating and controlling input/output operations,
4) handling errors and restarts,
5) coordinating communications between the human operator and the computer system,
6) maintaining a log of system operations and
7) controlling operations in a multiprogramming, multiprocessing, or time-sharing mode.

OPTIMUM START/STOP—An energy management application program which forecasts the optimum start time and stop time of the controlled plant or fan system in order to reduce equipment run-time while maintaining building comfort conditions at occupancy.

OR—A Boolean operator which is true when one or more of the inputs are true. Equivalent to a circuit of contacts arranged in parallel.

OS—See Operating System.

OUTPUT—The processed information being delivered by a computer. Computer results, such as a command or an English message.

OVERLAY—A popular technique for bringing routines into main memory from secondary memory during processing, so that several routines will occupy the same storage locations at different times. It is used when the total program memory requirements exceed the available main memory.

OVERLAY PROGRAM—A program that is executed or capable of being executed using the overlay technique.

OVERLAY SEGMENT—The smallest functional unit which can be loaded as one logical entity during execution of an overlay program.

P

PAGE—A segment of memory that is treated as a unit. For example, 65,536 bytes of memory (64K) might be divided into 16 pages of 4096 bytes. A page is usually between 512 and 4096 bytes or words.

PAGING—Segmenting programs into pages so that transfers between secondary and main memory can be made in pages rather than entire programs.

PAPER TAPE—See Tape, Paper.

PARAMETER—A quantity utilized by a program which may be given different values each time the program is executed but usually remains constant during a single execution of the program.

PARITY CHECK—The process of checking the parity bit to determine if a read error has occurred. See Bit, Parity.

PARTITION—A subdivision of main memory which is allocated to a system task for task execution. A partition may be fixed or variable in size.

PASCAL—A high-level language whose popularity is growing very rapidly. Many experts consider it to be the language of the 1980s. It was developed by Niklaus Wirth, a professor in Switzerland, whose objective was to simply develop a language that he could use in computer programming classes which had excellent structure and taught the students good programming habits. The language is named after Blaise Pascal (1623–1662), a French mathematician who built a successful digital calculating machine in 1642.

PASS—A scanning of source code by an assembler or a compiler. For example, a two-pass assembler is one that processes the source code in two separate steps.

PATCH—A section of coding inserted into a routine to correct a mistake or alter the routine. It is often not inserted into the actual sequence of the routine being corrected, but placed somewhere else, with an exit to the patch and a return to the routine provided. To insert such corrected code.

PC BOARD—Printed Circuit Board. A circuit board whose electrical connections are made through conductive material that is contained on the board itself, rather than with individual wires.

P-CODE—See Pseudo-code.

PCU—See Peripheral Control Unit.

PERIPHERAL—A device used for storing data, entering it into or retrieving it from the computer system.

PERIPHERAL CONTROL UNIT—An intermediary control device that links a peripheral to the central processor.

PERSONALITY CARD—An easily changeable electronic card within a PROM programmer. This card provides the correct timing patterns, voltage levels and other requirements to be met when programming a specific PROM chip. The card is changed when programming a different type of PROM chip.

PIXEL—Picture Element. A single point or dot in the grid of dots that forms the image on CRT display. See Graphic, Full and Graphic, Semi.

PLOTTER—A hard-copy peripheral device that produces line drawings such as X/Y graphs. The coordinates of the points or lines to be plotted are normally supplied by the computer.

PLUG-COMPATIBLE HARDWARE—Computers that can be operated with software originally developed for other kinds of computers.

PMOS—P-channel MOS. A MOS circuit in which holes are the charge carriers. Oldest type of MOS circuit.

POINT, CALCULATED—See Point, Pseudo.

POINTER—A memory location or register which contains a number that represents another main memory address. This is an indirect method of data retrieval.

POINT, PHYSICAL—A hardware device (sensor or actuator) that provides a discrete value or status as input to and/or output from a controller.

POINT, PSEUDO—A value or status created by a program and treated by the system as a physical point. Also called Calculated Point.

POLLING—A method used in multiprocessing to identify the source of interrupt requests. The interrogation of the devices in a system by the processor to determine if and where any I/O operations are pending.

POOL—A collection of interchangeable peripheral devices or main memory locations.

POPULATED BOARD—A circuit board that contains all of its electronic components. The components for an unpopulated board, conversely, must be supplied by the purchaser.

PORT—A communication channel between a computer and another device.

POWER DEMAND CONTROL—An energy management application program which controls electrical equipment in buildings in order to level the demand peaks that cause expensive surcharges.

POWER FAIL/RESTART—A facility that enables a computer to return to normal operation after a power failure.

PREPROCESSOR—A computer program used to add capabilities and features to a language or system. A FORTRAN preprocessor might extend the language to support structured programming techniques.

PRINTER, CHARACTER—A device that prints a single character at a time. Contrast with line printer. Speed is specified in characters per second.

PRINTER, LINE—An output device that prints an entire line of information at a time. Speed is specified in lines per minute.

POWER LINE CARRIER—A unit which allows a building's existing electrical wiring to transmit an EMS's signal.

PRIORITY—An order of precedence established for competing events.

PRIORITY, COMMAND—A priority level associated with a command sent to a system point. The command is executed if the command priority is greater than the residual priority of the last command executed on the point. See Priority, Residual.

PRIORITY, RESIDUAL—A priority level associated with a command sent to a point and which remains associated with that point if the command is executed. See Priority, Command.

PRIORITY SCHEDULING—Scheduling of programs and routines based on priority rather than position in queue.

PROCEDURE—The course of action taken for the solution of a problem.

PROCESSING, BACKGROUND—Low priority work done by the computer when real-time, conversational, and high priority (foreground) programs are inactive.

PROCESSING, BATCH—Processing of jobs which are submitted to run independently of events outside the system (as opposed to real-time or interactive jobs) and are normally processed on a deferred or time-independent basis.

PROCESSING, CONCURRENT—Execution of two or more programs simultaneously.

PROCESSING, DISTRIBUTED—A sharing of the processing (CPU) load by all computers in a distributed system. See Control, Distributed and Distributed Control System.

PROCESSING, FOREGROUND—Processing of programs in the foreground.

PROCESSING, IMAGE—Digitizing an image for subsequent image restoration, enhancement and information extraction by a digital computer.

PROCESSING, REAL-TIME—The processing of information or data in a sufficiently rapid manner so that the results of the processing are available in time to influence the process being monitored or controlled.

PROCESSING, REMOTE BATCH—The submission of jobs for batch processing from a remote terminal.

PROCESSOR—A device capable of receiving data, manipulating it, and supplying results, usually of an internally stored program. A program that assembles, compiles, interprets or translates. Also used synonymously with CPU.

PROCESSOR, DISPLAY—A processor, usually part of a graphics CRT terminal, which fetches information from a memory, where the picture to be displayed is defined, and converts it into the elementary driving signals for the analog circuits that generate the display or graphic.

PROCESSOR, FRONT-END—A small computer, often a microprocessor, that serves as a line or communications controller for a larger computer system.

PROGRAM—A set of coded instructions directing a computer to perform a particular function.

PROGRAM DEVELOPMENT SYSTEM—A hardware/software system supplied by a computer manufacturer to facilitate development and checkout of programs. System usually has a CRT terminal, floppy disk unit, hard copy printer and necessary software.

PROGRAM LIBRARY—See Library, Software.

PROGRAMMER—One who prepares programs for a computer.

PROGRAMMING, STRUCTURED—See Structured Programming.

PROM—Programmable Read-Only Memory. A type of semiconductor memory that is not programmed (encoded) during fabrication. Programming requires a later physical operation often performed by a PROM burner. Some PROMs can be erased after programming. See EPROM.

PROM BURNER—A hardware unit utilized to program instructions and data into a PROM memory chip.

PROM PROGRAMMER—See PROM burner.

PROTOCOL—A set of conventions between communicating processor governing the format and content of messages to be exchanged.

PSEUDOCODE—An instruction set for a nonexistent machine. Called P-Code. A mnemonic code which often looks like an assembly language. Because of the short life cycle of microprocessors, many modern day compilers generate a pseudocode output. A hardware or software interpreter then enables this code to be executed by several target micro-processors without rewriting the source program and without a new compiler for each machine.

PSEUDO-OP—Pseudo-operation. An instruction that is implemented by an assembler but is not in the processor's instruction set.

PSYCHROMETRIC ROUTINE—A utility routine which calculates five of the seven variables which describe the thermodynamic state of air when two of the variables are known.

R

RADIX—The number chosen to be the base of a number system. The decimal number system uses the radix 10.

RADIX POINT—The dot that indicates the separation between the integral and fractional parts of a number.

RAM—Random Access Memory. A type of memory whose access time is not significantly affected by the location of the data to the access. Usually implies volatile, semiconductor memory in modern microprocessor-based systems.

REAL-TIME—The performance of a computation during the actual time that the related physical process transpires in order that results of the computation can be used in guiding the physical process.

REAL-TIME CLOCK—A program-accessible clock which indicates the passage of actual time. The clock may be updated by hardware or software.

REAL-TIME EXECUTIVE—See Executive, Real Time.

REAL-TIME PROCESSING—See Processing, Real-Time.

RECORD—A collection of logically related fields of data.

RECURSIVE—Repetitive on a cyclical basis. Pertaining to a procedure which, while being executed, either calls itself or calls another procedure, which in turn calls the original one.

REENTRANT—An attribute of a program which allows one copy of that program in main memory to be used concurrently by two or more tasks. Attained by writing the program so that it does not modify itself and so that it maintains the current state of all concurrent executions.

REENTRANT ROUTINE—See Routine, Reentrant.

REGISTER—A memory location in the arithmetic and logic unit or in RAM used for the temporary storage of one or more words to facilitate arithmetic, logic or transfer operations. Special purpose registers are called the accumulator, address, index and instruction registers.

REMOTE BATCH PROCESSING—See Processing, Remote Batch.

RESIDENT—The property of being located in a memory device such as main memory or disk memory.

RESOURCE—Any facility of the computing system or operating system required by a job or task. These include main memory, input/output devices, the central processing unit, files, and control and processing programs.

RESOURCE ALLOCATION—Assigning a system resource to a program before or during execution.

RESPONSE TIME—The amount of time required for a computer to respond to an input from one of its peripherals.

RESTART, AUTOMATIC—The capability of a computer to perform automatically the initialization functions necessary to resume operation following an equipment or power failure.

ROLL-OUT/ROLL-IN—A method of increasing available main memory by temporarily storing an executing program on secondary storage, using the area occupied by the program, and then returning the program from secondary to main memory and restoring its status.

ROM—Read Only Memory. Memory whose preprogrammed contents cannot be altered by the computer. A nonvolatile, semiconductor memory. Almost every computer system has some ROM in which enough instructions data are stored to initialize the system after power is lost. See also PROM, EPROM and EAROM.

ROTATIONAL SHED/ADD—A power demand shed/add algorithm where loads are shed and added in a rotational order. Usually applied to loads of equal priority in the building complex.

ROUTINE—A set of instructions arranged in correct sequence that causes a computer to perform a particular process. Synonymous with program.

ROUTINE, DIAGNOSTIC—A utility routine which checks out a hardware device or helps locate a malfunction in the device.

ROUTINE, ERROR—A program which automatically initiates corrective action when a hardware or software error is detected.

ROUTINE, REENTRANT—A routine which does not alter itself during execution.

ROUTINE, SERVICE—A program or routine supporting the operation and maintenance of a computer system.

ROUTINES, FILE MANAGEMENT—The collection of operating system routines that are used to accomplish all the various file services provided by the system, such as file control, cataloging and file protection. Also called the system File Manager.

RPG—Report Program Generator Language. A high-level language designed specifically for writing programs which produce reports. RPG has been used widely for report generation in the business environment.

RS-232 AND RS-422—Technical specifications published by the Electronic Industries Association establishing the interface requirements between modems and terminals or computers.

RUN TIME—The time at which the program is executed. Also, the amount of time required to execute the program.

S

SCHEDULER—A program which schedules the execution of other programs. Some schedulers also initialize the scheduled program and allocate resources.

SCROLLING—New lines are added following the last line on a CRT screen. When the screen is full, all lines are moved up one line to make room for the new line at the bottom. The top line is then lost.

SECOND SOURCE—A manufacturer who produces a product that is interchangeable with the product of another manufacturer.

SECTOR—Addressable storage unit on a magnetic disk. Disk tracks (concentric bands) are divided into sectors each having the same capacity for storage of data.

SEGMENT—A portion of a large program small enough to fit into main memory for execution. See Overlay and Overlay Segment.

SEMICONDUCTOR—A material such as silicon with a conductivity between that of a metal and an insulator. It is used in the manufacture of solid-state devices such as diodes, transistors and the complex integrated circuits that comprise computer logic circuits.

SEQUENTIAL SHED/ADD—A power demand shed/add algorithm which sheds and adds electrical loads based on a fixed priority sequence. First shed, last add philosophy.

SERVICE ROUTINE—See Routine, Service.

SETPOINT—The required or ideal value of a controlled variable.

SHED ORDER—A priority level assigned to electrical loads controlled by a Power Demand Control Program.

SIMULATION—A computer program that mathematically models a process.

SIMULATOR—A utility program that runs on one computer and imitates the operations of another computer. A simulator allows software development to proceed independently of associated hardware development. In fact, hardware need not even exist.

SIMULATOR, SIP—A program that provides the same functionality as the SIP.

SINGLE STEPPING—Having the computer execute a program slowly so that the user can watch each step and its effects.

SLAVE—An element of a computing system that is under the functional control of a similar element.

SLIDING WINDOW—A power demand control algorithm which controls power consumption over a demand interval which is continuously indexed ahead one sample period at the end of each sample period.

SNOBOL—StriNg Oriented symBOLic language. A high-level language developed to facilitate the manipulation of strings of characters. SNOBOL is particularly suited for text editing, linguistics and symbolic manipulation of algebraic expressions.

SOFT COPY—Output presented on a video display.

SOFT SECTORING—Defining the sector format of a disk through software.

SOFTWARE—A collection of programs and routines associated with a digital computer which make the computer perform a scientific function or which facilitate the programming of that computer. A change in software can make the same computer perform a completely different function. Software includes operating systems, compilers, language processors, utility routines, and application programs.

SOFTWARE, APPLICATION—Software which has value to a user in reducing operating costs or increasing system performance but which is optional in that it is not required to make any hardware unit perform its basic function.

SOFTWARE, CANNED—Off-the-shelf, proven programs designed to perform one or more general functions.

SOFTWARE DESIGN METHODOLOGY—A set of work procedures for performing the requirements analysis and program design. A scientific discipline for systematically deriving specifications and the algorithms which make up the software design.

SOFTWARE SUPPORT—A service offered, for an annual fee, which entitles a software customer assistance by telephone and mail, a subscription to software release bulletins, and a subscription to all new releases of the software products covered under the support agreement. Software support does not include any on-site services. Synonymous with software maintenance.

SOFTWARE, SYSTEM—Software which is required to make a hardware unit perform its basic function. Also called base software.

SOFTWARE TOOLS—A collection of notations, programs and management procedures which result in productive software development.

SOFTWARE, UTILITY—Software which facilitates the installation, start-up, checkout, troubleshooting, maintenance, and manipulation of other software.

SOS—Silicon on Sapphire. Refers to the layers of material and, indirectly, to the process of fabrication of devices that achieve bipolar speeds through MOS technology by insulating the circuit components from each other.

SOURCE CODE—See Code, Source.

SOURCE PROGRAM—A computer program as written in source code by the programmer. It must be translated into an object program before execution by the computer.

STACK—A sequential data list stored in main memory. Rather than addressing the stack elements by their memory location, the processor retrieves information from the stack by popping elements from the top or from the bottom.

START/STOP PROGRAM—A special type of Event Initiated Program which automatically initiates a sequence of commands to building equipment or points at specified times.

STARTUP—The process of loading and initializing an operating system.

STATEMENT—A single computer instruction within a computer program.

STORAGE—See Memory.

STORAGE, BULK—Same as Storage, Mass.

STORAGE, MASS—Auxiliary or bulk memory as opposed to main memory. Disk drives and tape drives are common mass storage devices. Synonymous with Secondary Memory or Bulk Memory.

STRING—A set of contiguous characters treated as a unit by the computer.

STRUCTURED PROGRAMMING—A term encompassing a broad range of programming techniques and disciplines all directed toward the objective of making software development more productive and more of a science.

SUBPROGRAM—A part of a larger program that can be compiled independently. Often used synonymously with subroutine and routine.

SUBROUTINE—A routine which can be called (used) by another routine to perform a specific function after which control is returned to the calling routine. Compare with function.

SUPERVISOR—See Executive.

SUSPENDED—The state of a program whose execution has been halted pending the occurrence of some specified event.

SWAPPING—A method for sharing main memory between several programs by maintaining each program and its status on secondary memory and loading each one into main memory for a limited time interval.

SYMBOLIC LOGIC—See Boolean Algebra.

SYNCHRONOUS—Occurring concurrently and with a regular or predictable time relationship. Opposite of asynchronous.

SYNCHRONOUS COMMUNICATION—Data transmission where the bits are transmitted at a fixed rate. The transmitter and receiver both use the same clock signals for synchronization.

SYNTAX—The structure of expressions in a language. The rules governing the structure of a language.

SYSTEM—The computer and all its related components.

SYSTEM, SPECIAL—A nonstandard hardware and/or software configuration designed, developed and installed for a unique customer request.

SYSTEM, STANDARD—A system configured and installed using off-the-shelf hardware and software.

T

T AND V ROUTINES—Test and Verify Routines. Utility programs to check out hardware devices and to diagnose and locate hardware faults.

TAPE, CASSETTE—A secondary memory device employing the principles of a magnetic tape where the tape is stored permanently in a cassette. One cassette is capable of storing over one million bits of information. While access time is slow, typically from 10 seconds to several minutes, this is a very inexpensive and portable storage medium.

TAPE DRIVE—A device that moves tape past a head.

TAPE, MAGNETIC—A plastic tape coated with a magnetic film on which data and instructions can be stored. The information can be stored digitally as a sequence of magnetized dots on the cassette tape surface. The information can also be stored as sounds. One tone represents a zero while another tone represents a one. The audio method is least expensive and is typically used on cassette tapes.

TAPE, PAPER—A length of narrow paper punched in a pattern decodable by a paper tape reader. Paper tape is bulkier and slower than magnetic forms of data storage. On the other hand, paper tape cannot be erased accidentally and is easier to edit by splicing.

TAPE, 9-TRACK—A secondary memory unit. The magnetic tape unit is called "9-track" because 9 bits of information are stored in parallel across the width of the tape. The tape is a ½ inch wide, plastic tape and a separate read-write head is provided for each bit positioned across the tape. 8 of the bits represent one byte of information while the 9th bit is used as a check or parity bit. Magnetic tape storage is nonvolatile. A single 10½-inch diameter reel of tape stores over 23 million 8-bit words.

TARGET COMPUTER—The computer on which the software under development will execute operationally.

TASK—A program, routine or subroutine. A program may be broken into several smaller programs or tasks.

TELECOMMUNICATION—A general term expressing data transmission between a computing system and remotely located devices via a unit which performs the necessary format conversion and controls the rate of transmission.

TELEPROCESSING—Same as telecommunication.

TEMPLATE—A display presented on a CRT screen which guides the operator in entering data by giving the name of the data to be entered and the required format of each entry.

TERMINAL—A device for communication with a computer. A typical terminal consists of a keyboard and a printer or video display.

TEXT EDITOR—See Editor.

THROUGHPUT—A measure of the amount of work that can be accomplished by the computer during a given period of time.

TIME SHARING—The simultaneous utilization of a computer system from multiple terminals. Performing several independent processes almost simultaneously by interleaving the operations of the processes on a single processor.

TRACE—A debugging technique which provides an analysis of each executed instruction and writes it on an output device as each instruction is executed.

TRACE PROGRAM—A computer program that records the chronological sequence of events taken by another program during its execution.

TRANSLATOR—A program whose input is a sequence of statements in some language and whose output is an equivalent sequence of statements in another language.

TRANSPARENT—The property of being invisible to the user, another program or a hardware device.

TRAP—A hardware or software implemented function that signals the processor whenever a specified condition occurs.

TTL—Transistor-Transistor Logic. A very fast type of bipolar logic circuit which consumes more power than CMOS. Takes its name from the method of interconnecting transistor components.

TROUBLESHOOT—Same as Debug.

TTY—Teletypewriter. A hard-copy terminal.

TURNKEY—A product delivered ready to run.

U

UART—Universal Asynchronous Receiver-Transmitter. An integrated circuit that converts parallel input into serial form, or vice versa.

UNBUNDLED—Separate pricing of hardware and software. For many years, computer manufacturers included software, without additional charge, when their computer hardware was purchased. IBM began to unbundle its software in the late 1960s and most of the computer industry has followed thus opening a giant software market and creating the "software house."

UNCONDITIONAL BRANCH—See Jump.

UNCONDITIONAL JUMP— See Jump.

UTILITY—A program which facilitates the installation, start-up, checkout, troubleshooting, maintenance, and manipulation of other software.

V

VIRTUAL—Apparent or imaginary as contrasted to actual or absolute.

VIRTUAL DEVICE—An imaginary device that the processor assumes to be present.

VIRTUAL KEY—A key on a computer input device whose function can be changed by software to fit the current operational need.

VIRTUAL MEMORY—See Memory, Virtual.

VLSI—Very Large Scale Integration. The process of integrating over 100,000 circuits or equivalent components on a single chip of semiconductor material such as silicon. More than 2000 logic gates or 16,000 bits on a chip.

W

WALKTHROUGH—A meeting during program development, at which a programmer explains the logic of his program to his peers. The objective is to expose logic faults before the program is written (coded).

WATCHDOG TIMER—A timer set by a program to prevent the system from looping endlessly or becoming idle because of program errors or hardware faults.

WORD—A set of binary digits (bits) that occupies one storage location in memory and is treated by the computer circuits as a unit and transported as such.

WORD LENGTH—The number of bits in a computer word.

WRITE—To transfer information, usually from main storage to an output device.

Appendix B

Abbreviations

AA	Analog Alarm
ac	Alternating Current
A/D	Analog to Digital
AEE	Association of Energy Engineers
AI	Analog Input
ALGOL	ALGOrithmic Language
ALU	Arithmetic and Logic Unit
ANSI	American National Standards Institute
AO	Analog Output
APEM	Association of Professional Energy Managers
ASCII	American Standard Code for Information Interchange
ASHRAE	American Society of Heating, Refrigeration, and Air Conditioning Engineers
ATC	Automatic Temperature Control
BASIC	Beginners' All-Purpose Symbolic Instruction Code
BCD	Binary Coded Decimal
bps	Bits Per Second
CCC	Central Communications Controller
CCDs	Charge Coupled Devices
CCU	Central Control Unit
CIL	Control Interpreter Language
CLM	Command Line Mnemonic
CLMI	Command Line Mnemonic Interpreter
CLT	Communications Link Termination
COBOL	Common Business Oriented Language

CPA	Control Point Adjustment
cps	Characters Per Second
CPU	Central Processing Unit
CRT	Cathode Ray Tube
CT	Current Transformer
D/A	Digital to Analog
dB	Decibel
DBL	Database Language
dc	Direct Current
DDC	Direct Digital Control
DDL	Data Description Language
DE	Data Environment
DGP	Data Gathering Panel
DI	Digital Input
DIP	Dual In-line Package
DMA	Direct Memory Access
DMBS	Database Management System
DML	Data Manipulation Language
DO	Digital Output
DOS	Disk Operating System
DTC	Data Terminal Cabinet
DTM	Data Transmission Media
DX	Direct Expansion
EAROM	Electrically Alterable ROM
EBCDIC	Extended Binary Coded Decimal Interchange Code
EDAC	Error Detection and Correction
EDP	Electronic Data Processing
EIP	Event Initiated Program
EMCS	Energy Management and Control Systems
EMI	Electromagnetic Interference
EEPROM	Electrically Erasable Programmable Read Only Memory

EPROM	Erasable PROM	**PROM**	Programmable Read-Only
FCB	Failure Control Board		Memory
FID	Field Interface Device	**RAM**	Random Access Memory
FORTRAN	FORmula TRANslator	**RF**	Radio Frequency
HIPO	Hierarchical Input-Process-	**RFI**	Radio Frequency
	Output		Interference
IC	Integrated Circuit	**ROM**	Read Only Memory
I/O	Input/Output	**RT**	Run Time
LSI	Large Scale Integration	**RTC**	Real Time Clock
MA	Milliamp	**RTD**	Resistance Temperature
Mb	Megabyte		Detector
MCR	Master Control Room	**SCR**	Silicon Controlled Rectifier
MHz	Megahertz	**S/N**	Signal to Noise Ratio
MLCP	Multiline Communications	**SNOBOL**	StriNg Oriented SymBOLic
	Processor		Language
MMI	Man-Machine Interface	**S/S**	Start/Stop
MODEM	Modulator/Demodulator	**TTL**	Transistor-Transistor
MOS	Metal Oxide Semiconductor	**TTY**	Teletypewriter
MPU	Microprocessor Unit	**UART**	Universal Asynchronous
MUX	Multiplexer		Receiver-Transmitter
NS	Nanosecond	**VLSI**	Very Large Scale Integration
PCU	Peripheral Control Unit		

Appendix C

Suppliers of Energy Management Systems – Building Controls and Automation

ADT Security, 2 World Trade Center, New York, N.Y. 10048

AET Systems, 77 Accord Park Drive, Norwell, Mass. 02061

Advanced Micro Systems, 9076 N. Deerbrook Trail, Milwaukee, Wisc. 53223

Aegis Systems Corp., 607 Airport Blvd., Doylestown, Pa. 19801

American Air Filter, 214 Central Ave., Louisville, Ky. 40208

American Auto-Matrix Inc., 1 Technology Drive, Export, Pa. 15632

American Multiplex Systems Inc., 1148 East Elm Avenue, Fullerton, Calif. 92631

Anderson Cornelius Co., 6750 Shady Oak Road, Eden Prairie, Minn. 55344

Andover Controls Corp., York and Haverhill Sts., Andover, Mass. 01810

Automated Logic Corp., 120 Interstate North Parkway East, Suite 436, Atlanta, Ga. 30339

Barber-Colman Co., Environmental Controls Div., 1354 Clifford Ave., Loves Park, Ill. 61132

Barrington Associates, Suite 200, 700 El Camino Real, Millbrae, Calif. 94030

Bion Control Corp. of America, 2410 Gilbert Ave., Cincinnati, Ohio 45206

Borg Warner Air Conditioning, P.O. Box 1592, York, Pa. 15218

Butler Controls Div., Butler Mfg. Co., P.O. Box 2249, Redmond, Wash. 98033

CETEK Systems, 1620 C. Berryessa Road, San Jose, Calif. 95133

Computer Controls Corp., 54 Industrial Way, Wilmington, Mass. 01887

Conduff-Rogers, P.O. Box 13, Springdale, Ark. 72765

Control Pac Corp., 23840 Industrial Park Drive, Farmington Hills, Mich. 48024

Control Systems International, P.O. Box 59469, Dallas, Tex. 75229

Controlled Energy Systems Co. (Cesco), 1240 N.E. 175th St., Seattle, Wash. 98155

Dencor Inc., 2750 S. Shoshone, Englewood, Colo. 80110

Detection Systems Inc., 130 Perinton Parkway, Fairport, N.Y. 14450

EDA Controls Corp., 6645 Singletree Drive, Columbus, Ohio 43229

EDPAC Corp., 200 Welsh Road, Horsham, Pa. 19044

EIL Instruments, 10 Loveton Circle, Sparks, Md. 21152

Elan Energy Systems, 2562 Middlefield Road, Mountainview, Calif. 94043

Electronic Systems USA, 1014 East Broadway, Louisville, Ky. 40204

Encon Systems Inc., 502-F Van Dell Way, Campbell, Calif. 95014

Enercon Data Corp., 7464 West 78 St., Minneapolis, Minn. 55435

Energy Management Corp., 9 Shilling Road, Hunt Valley, Md. 21031

Engineered Energy Systems, 110 Dorsa Ave., Livingston, N.J. 07039

Fluidmaster, 1800 Via Burton, Anaheim, Calif. 92806

Functional Devices Inc., 310 S. Union St., Russiaville, Ind. 46979

General Electric Co., Wiring Device Dept., 225 Service Ave., Warwick, R.I. 02886

General Electronic Engineering Inc., 132 West Main St., Rahway, N.J. 07065

Heat-Timer Corp., 10 Dwight Place, Fairfield, N.J. 07706

Honeywell Building Controls Div., 1985 Douglas Drive N., Golden Valley, Minn. 55422-3992

Honeywell Building Systems Div., Honeywell Plaza, Minneapolis, Minn. 55408

Honeywell Building Systems Div., Albuquerque Systems, 8500 Bluewater N.W., Albuquerque, N.M. 87105

HSQ Technology, 1435 Huntington Ave., South San Francisco, Calif. 94083-2248

Hypertek Inc., P.O. Box 137, Whitehouse, N.J. 08888

IBM-Automation Systems Division, 3715 Northside Parkway, Atlanta, Ga. 30055

IDMA Synergy/Climatron, 11362 Western Ave., Stanton, Calif. 90680

Intermatic Inc., Intermatic Plaza, Spring Grove, Ill. 60081

Insyte Inc., 1158 Lenor Way, San Jose, Calif. 95128

Jade Controls, P.O. Box 271, Montclair, Calif. 91763

Johnson Controls Inc., 507 E. Michigan St., Milwaukee, Wisc. 53201

Lumenite Electronic Co., 2331 N. 17th Ave., Franklin Park, Ill. 60131

MCC Powers, Unit of Mark Controls Corp., 2942 MacArthur Blvd., Northbrook, Ill. 60062

MicroControl Systems Inc., 6579 N. Sidney Plaza, Milwaukee, Wisc. 53209

Nova Energy Systems, 39 Washington Ave., Point Richmond, Calif. 94801

Novar Controls Corp., 24 Brown St., Barberton, Ohio 44203

Paragon Electric Co., 606 Parkway Blvd., P.O. Box 28, Two Rivers, Wisc. 54214

PowerLine Communications Inc., 123 Industrial Ave., Williston, Vt. 05495

Robertshaw Controls, Control Systems Div., 1800 Glenside Drive, Richmond, Va. 23226

Robertshaw Controls Co., Integrated Systems Div., 3000 D South Highland Drive, Las Vegas, Nev. 89109

Robertshaw Controls Co., Uni-Line Div., P.O. Box 2000, Corona, Calif. 91720

Sachs Energy Management Systems, Inc., P.O. Box 96, St. Louis, Mo. 63166

Scientific Atlanta Inc., P.O. Box 105038, Atlanta, Ga. 30348

Service Master Inc., 2300 Warrensville Road, Downers Grove, Ill. 60515-1725

Signaline, Div. of Time Mark Corp., 11440 E. Pine, Tulsa, Okla. 74116

Softmatic Control Systems Inc., 16935 W. Bernardo Drive, Rancho Bernardo, Calif. 92128

Solid State Systems Inc., 1990 Delk Industrial Blvd., Marietta, Ga 30067

Solidyne Corp., 2207 Hammond Drive, Schaumburg, Ill. 60195

Staefa Control Systems Inc., 8515 Miralani Drive, San Diego, Calif. 92126

T. J. Controls Inc., 10130 S.W. Nimbus, D5, Portland, Ore. 97223

TSBA Controls Inc., 52-35 Barnett Ave., Long Island City, N.Y. 11104

Tech-S Inc., 12997 Merriman Road, Livonia, Mich. 48150

Teletrol System Inc., 340 Commercial St., Manchester, N.H. 03101

Toledo Scale, Div. of Reliance Electric, 350 W. Wilson Bridge Road, Worthington, Ohio 43085

Tork Inc., 1 Grove St., Mt. Vernon, N.Y. 10550

Tour & Anderson Inc., 652 Glenbrook Road, P.O. Box 2337, Stamford, Conn. 06906

The Trane Co., 3600 Pammel Creek Road, La Crosse, Wisc. 54601

The Trane Co., Building Automation Systems Division, 5301 East River Drive, Minneapolis, Minn. 55412-1096

Triangle MicroSystems Inc., 8600 Jersey Court, Raleigh, N.C. 27612

Tropic Kool Engineering Corp., 12900 Automobile Blvd., Clearwater, Fla. 33520

TWR Associates, 515 Oakland Ave., Oakland, Calif. 94611

United Technologies Building Systems Co., 10 Farm Springs, Farmington, Conn. 06032

Westinghouse Electric Co., Integrated Building Systems, Parkway Center, Building 5, Pittsburgh, Pa. 15220

The above list should not be construed as complete. The list of energy management systems suppliers includes only companies that put their names on the equipment they sell. It does not include distributors or dealers.

Index